Get on the Air with HF Digital

Steve Ford, WB8IMY

Printed in USA

ISBN: 978-1-62595-159-5

Third Edition
Second Printing

Contents

About ARRL

We're the American Radio Relay League, Inc. — better known as ARRL. We're the largest membership association for the amateur radio hobby and service in the US. For over 100 years, we have been the primary source of information about amateur radio, offering a variety of benefits and services to our members, as well as the larger amateur radio community. We publish books on amateur radio, as well as four magazines covering a variety of radio communication interests. In addition, we provide technical advice and assistance to amateur radio enthusiasts, support several education programs, and sponsor a variety of operating events.

One of the primary benefits we offer to the ham radio community is in representing the interests of amateur radio operators before federal regulatory bodies advocating for meaningful access to the radio spectrum. ARRL also serves as the international secretariat of the International Amateur Radio Union, which performs a similar role internationally, advocating for amateur radio interests before the International Telecommunication Union and the World Radiocommunication Conference.

Today, we proudly serve nearly 160,000 members, both in the US and internationally, through our national headquarters and flagship amateur radio station, W1AW, in Newington, Connecticut. Every year we welcome thousands of new licensees to our membership, and we hope you will join us. Let us be a part of your amateur radio journey. Visit www.arrl.org/join for more information.

225 Main Street
Newington, CT 06111-1400 USA
Tel: 860-594-0200
FAX: 860-594-0259
Email: membership@arrl.org
www.arrl.org

Chapter 1

Station Hardware and Software

The HF Transceiver

The radio requirements for HF digital are surprisingly straightforward. You don't need to run out and purchase a special radio with sophisticated features. All you need is an SSB voice transceiver.

Any transceiver made within the last 20 years will work well with most HF digital modes. When considering an older radio, however, there are two things to keep in mind:

Stability — Older transceivers may tend to drift in frequency. Drift is deadly to digital communications. If you must use an old rig as your HF digital transceiver, you may need to allow it to warm up for as long as 30 minutes prior to operating.

Transmit/Receive Switching Speed — One digital mode known as PACTOR requires the radio to jump from receive to transmit (and back) in milliseconds. Many older radios can achieve this, but it is hard on their switching relays. Newer radios with solid-state switching are better for this application.

If your radio dates from about 1990 to the present day, you'll be in fine shape. These rigs are stable and most include all the features you'll need for HF digital.

Radio Ports

Modern radios get along surprisingly well with computers and other external devices. In fact, most modern HF transceivers are designed with external devices in mind. They offer a variety of ports depending on the model in question.

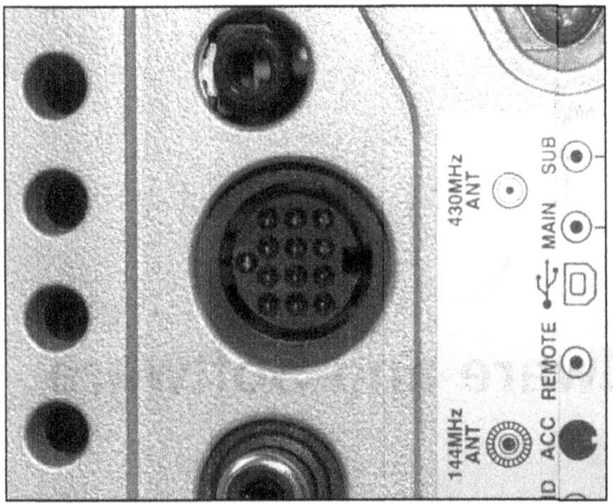

Figure 1.1 — A typical 13 pin transceiver accessory jack. Among the pins of most interest to HF digital operators are those that carry receive audio, transmit audio and the PTT (Push to Talk) line.

Nearly every HF rig manufactured within the past two decades includes an "accessory" port of some kind. Typically, these are multipin jacks (as many as 13 pins) that provide connections for audio into and out of the radio, as well as a pin that causes the radio to switch from receive to transmit whenever the pin is grounded. This is often called the *PTT* or *Push to Talk* line. Some manufacturers also call it the "Send" line. Look at the typical accessory jack shown in **Figure 1.1**.

These accessory ports are ideal for connecting the kinds of interface devices we use to operate HF digital. In addition to the PTT function, accessory ports provide receive audio output at *fixed* levels that never change no matter where the VOLUME knob is set. This is a highly convenient feature that you'll appreciate when operating late at night after everyone has gone to bed. You can turn the VOLUME knob to zero and still have all the receive audio you need!

Be aware that some radio manufacturers label accessory ports as "data" or "digital" ports. This causes no end of confusion because modern rigs often offer two types of connections: a true accessory port with audio and transmit/receive keying lines and another port that allows a computer to actually control the radio. The confusion occurs when hams attempt to figure out which ports they need to use.

For the purpose of getting on the air with HF digital, the only type of radio "control" we care about is the ability to switch from receive to transmit and back again. That connection is available at the *accessory port*, even though the port may go by a different name.

The other type of control the transceiver manufacturers have in mind goes way beyond the act of simply switching between transmit and receive. They are talking about the computer taking over almost every function of the radio; that's a different animal entirely. Full computer control usually involves software that does many things such as displaying and changing the transceiver's frequency, raising and lowering the audio level, scanning memory frequencies and a great deal more. Some computer control software is so elaborate that the radio itself can be placed out of sight and all control conducted at the keyboard and monitor screen. Many amateurs use this capability to control their rigs remotely over the internet at great distances.

Transceivers have separate ports for this type of computer interfacing and these are most definitely *not* accessory ports. On the contrary, they are ports strictly designed to swap data with external computers. They come in several varieties …

TTL: Transistor-Transistor Logic. These ports require a special interface to translate the serial communication from your computer to TTL pulses that your radio can comprehend.

USB: Universal Serial Bus. Although the consumer electronics world adopted USB years ago, transceiver manufacturers have been somewhat slower to catch on.

RS232: This is a serial port that can be connected directly to your computer if your computer has a serial (COM) port. Most modern computers have done away with serial ports, but you can use a USB-to-serial converter to bridge the gap.

Ethernet: This port allows the transceiver to become a "network device," just like your wireless router, printer, etc. Only a handful of transceivers offer Ethernet ports currently.

If your transceiver lacks an accessory port, you can still connect your interface to the microphone and headphone jacks.

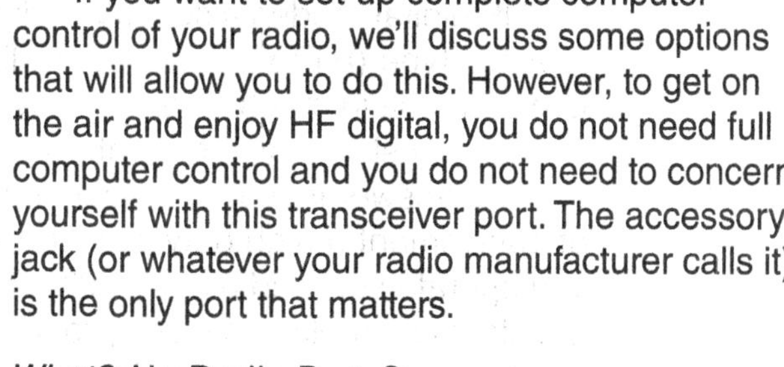

If you want to set up complete computer control of your radio, we'll discuss some options that will allow you to do this. However, to get on the air and enjoy HF digital, you do not need full computer control and you do not need to concern yourself with this transceiver port. The accessory jack (or whatever your radio manufacturer calls it) is the only port that matters.

What? No Radio Ports?

What if your SSB transceiver doesn't have an accessory port? No problem.

You can use the microphone jack as your connection for transmit/receive switching as well as the audio input. For the audio output, you can use the external speaker or headphone jack. This isn't an ideal situation, but it works. In fact, many HF digital operators take this approach.

Duty Cycle

When talking about your transceiver, we need to spend a little time discussing the concept of *duty cycle*. A somewhat crude definition of duty cycle is the time that a radio spends generating RF output as a fraction of the total time under consideration. In HF digital terms, think of duty cycle as the amount of time your radio is generating RF during any given transmission compared to the amount of the time during the same transmission when RF output falls to zero. A 100% duty cycle would mean that your radio is cranking out RF continuously throughout the entire transmission; the RF output level never falls to zero.

When you are transmitting digital, CW or even SSB voice, your rig may not be operating at a 100% duty cycle. Consider SSB voice as an example. Whenever you speak into the microphone, the RF output level changes dramatically as your voice changes. It can go from 100% output to zero in a fraction of a second. The same is true for CW. Whenever your CW key is open between the dots and dashes, your transceiver output is at zero.

Measured over a period of time (your transmitting time), the duty cycle of SSB voice is about 40%; CW is often as low as 30% or even less if you are a particularly slow sender.

HF digital modes also vary in duty cycle. Some modes such as radioteletype (RTTY) push your radio to a duty cycle of nearly 100%. Others result in much lower duty cycles.

So why should you care about your duty cycle?

The answer is that your radio may not be designed for the type of punishment a high duty cycle transmission can inflict. When you operate your radio at a 100% duty cycle, you are demanding that its final amplifier circuits produce the full measure of output — whatever you've set that output level to be — for the entire time you are transmitting. The result is heat, and potentially a lot of it. Apply enough heat to a circuit for a sufficient length of time and you'll see components begin to fail, sometimes in spectacular fashion.

Some manufacturers don't consider the possibility that their SSB voice transceivers might be pressed into service as digital transceivers. They design the radios to tolerate the duty cycle of a typical voice transmission. If you use this same radio to enjoy a high duty cycle digital mode such as RTTY, you could be asking the radio to operate well outside its design limits — with unfortunate results.

Always read your transceiver manual before attempting to use the radio for digital operating. The manufacturer may advise you to reduce the RF output by as much as 50% when using high duty cycle modes. This keeps the heat generation manageable. When in doubt, or when you notice that your radio is becoming particularly hot, reduce the RF output. You'll find that digital modes don't require a great deal of output power anyway, so chances are you won't notice the 50% reduction.

Transceiver Filters

Another important transceiver feature to consider is receive filtering. In most cases we're talking about the filters located in the Intermediate Frequency (IF) stage of your radio. These can be physical filters; modules that plug in or are soldered into the radio's circuit board. Many modern transceivers use Digital Signal Processing (DSP) rather than physical filters, this is particularly true of software-defined transceivers. The advantage of DSP is that it is often designed to be continuously variable. This means that you can narrow or expand the filter with the push of a button or the twist of a knob.

As you'll learn later, most popular HF digital modes use wide receive-audio bandwidths for reception, so it isn't strictly necessary to have a filter narrower than the typical SSB voice bandwidth of about 2.8 kHz. All SSB transceivers meet this requirement.

There are some important exceptions, though. If you decide to try your hand at RTTY contest operating, you'll quickly discover that a 2.8 kHz IF bandwidth is entirely too wide. You won't be able to easily separate individual signals in a sea of RTTY contest chaos. In this environment you need to narrow the IF bandwidth to at least 500 Hz. In ultra-crowded conditions you may need a bandwidth as narrow as 300 Hz.

There may be other situations where you'll want to use a narrower IF bandwidth. Let's say you are trying to communicate with a weak station and a much stronger station begins transmitting a rock-crushing signal about 1 kHz up the band. Your radio's Automatic Gain Control (AGC) is going to respond by dropping your receiver sensitivity into the basement. The weak station will become weaker still or disappear altogether.

It isn't practical to ask the strong station to move, so your only alternative is to dramatically narrow your IF bandwidth to the point where his signal is eliminated and only the weak signal remains. Continuously adjustable IF DSP filtering is terrific for this application.

You can create an IF filter as narrow as necessary and put it right on the signal you are trying to hear. With ultra-sharp DSP filters, everything outside the passband is effectively gone.

AFSK vs FSK

Here is an interesting point that sometimes causes consternation among digital operators when they consider their HF transceivers.

Basic HF SSB transceivers typically offer at least three operating modes: Upper Sideband (USB), Lower Sideband (LSB) and CW. Others include AM and even FM. For HF digital operation USB or LSB is all you need.

That said, you'll also notice that some HF rigs include a digital mode selection that may be labeled "RTTY," "Digital" or "Data." Here is where the gnashing of teeth begins because the way the radio behaves when you select this mode can vary depending on the design. In other words, these labels can mean different things in different radios.

For instance, the "Digital" mode may include a function that "looks" for the incoming audio at the accessory jack exclusively. Otherwise, it ignores the incoming audio at the accessory jack entirely.

What is most important for our discussion, however, is the fact that some manufacturers use this mode label to indicate that the radio will switch to *FSK*. What does this mean?

Most HF digital operators use Audio Frequency Shift Keying or *AFSK*, although they may not realize it. The audio tones from their sound devices are applied to their radios and converted to shifting RF frequencies at the output, hence the term *Audio* Frequency Shift Keying.

But there is yet another way to create shifting RF. You can take data pulses directly from the computer (not sound) and use those pulses to directly shift a transceiver's master oscillator from one RF frequency to another. Since there is no audio involved, this is known as Frequency Shift Keying or *FSK*.

Is there an advantage to using FSK vs AFSK? Not really. It is mostly a matter of convenience. When operating FSK, you never have to worry about pumping too much audio into the transceiver and causing distortion. Also, some older HF digital transceivers will not allow you to use narrow IF filters unless you operate using the FSK mode.

The problem with FSK in amateur transceivers is that it is limited to shifting between only two frequencies. Many amateur digital modes use more than two frequencies. The only common digital mode that can benefit from true FSK is RTTY since RTTY signals only shift between two frequencies.

FSK as it exists in amateur radio transceivers is a feature for RTTY aficionados alone. The irony is that it is impossible to tell the difference between a properly modulated AFSK RTTY signal and a RTTY signal generated by FSK.

If your transceiver offers a "RTTY," "Data" or "Digital" mode, read your manual carefully and find out exactly what this means. Chances are that "RTTY" means FSK. "Data" or "Digital" likely means AFSK, although even the manufacturers confuse the terminology. You'll find some manuals referring to AFSK as FSK. The authors are not wrong, strictly speaking. After all, regardless of whether you're using AFSK or FSK the RF signal frequencies at the output are still shifting back and forth. It is "frequency shift keying" in either case. The difference is in method you use to create the shifty signals.

If you're unsure, look at the hookup diagram in the manual. Does it show digital transmit audio being applied to the radio at the accessory or microphone jack? If so, it is AFSK. When you select the "Digital" or "Data" mode, you're really in the SSB mode, but using audio to generate the frequency changes.

But if the manual shows keying data (not audio) being supplied to an "FSK" line, it is indeed true FSK.

Computers

A desktop computer is housed in a separate, stand-alone case connected to a monitor screen, keyboard and mouse. Inside the computer case there is a sound device of some sort, either a dedicated *sound card* or a set of sound-processing chips on the motherboard. The computer connects to peripheral devices using USB (COM) ports. Desktop computers are still quite popular in amateur radio stations today.

Desktop PCs remain popular in amateur radio stations, although larger motherboard and cabinet designs are giving way to much more compact models like this one.

In many ham stations, laptops are taking over from desktop computers. They have become particularly popular for portable operating.

In recent years, laptop computers have challenged desktop machines for dominance. Modern laptop computers are every bit as powerful as many desktop models, yet they are less complicated and more convenient to use. This is especially true for portable operating. The good news is that most HF digital software does not require powerful computers. Any ordinary off-the-shelf consumer-grade computer will do the job. If you are buying new, don't overspend for a powerful computer you won't need.

In terms of operating systems, the vast majority of HF digital software is written for Microsoft *Windows*. Some of these programs were created during the *Windows* XP era, but they run well on both *Windows* 10 and 11.

There is HF digital software available for *MacOS* as well, although not as much variety. *Linux* users will find a number of HF digital applications, too.

If you are considering a computer purchase (desktop or laptop), here are a few rule-of-thumb shopping specifications:

• The processor clock frequency ("speed") should be 2 GHz or better.

• A hard drive (solid state or electromechanical) with at least 250 GB capacity.

• The computer should have as much memory as possible, not just to run the ham applications smoothly, but the operating system as well. For *Windows* 10 or 11, I'd recommend at least 8 GB of RAM; more is always better.

• Either a built-in wireless (Wi-Fi) modem or an Ethernet port. Although it isn't necessary for ham work, chances are you'll want to connect your station to the Internet from time to time. If so, you'll need a wireless modem or Ethernet port to do so.

The Importance of Sound

With few exceptions, every computer you are likely to purchase today will include a either a dedicated sound card or a sound chipset. This feature is critical for HF digital operation because most of the modes you'll enjoy depend on sound devices to function as radio modems — modulators/demodulators.

The audio from your radio enters your computer via the sound device where it is converted (demodulated) to digital data for processing by your software. The results are words or images on your computer monitor. When you want to transmit, this same sound device takes the data from your software, such as the words you are typing, and converts it to shifting audio tones according to whatever mode you are using. This conversion is a form of modulation. The tones are then applied to your radio for transmission.

The simplest built-in sound devices are those found in laptops. They provide two ports to the outside world: microphone (audio input) and headphone (audio output). In modern laptops these are

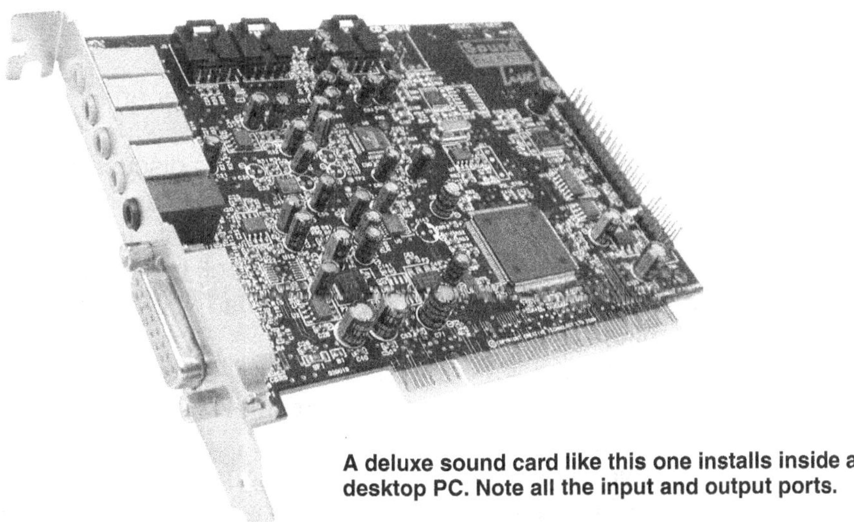

A deluxe sound card like this one installs inside a desktop PC. Note all the input and output ports.

often combined into one multiconductor port. These are perfectly adequate for HF digital work. Desktop computers often have a similar arrangement, although the output port is usually labeled "speaker." A "line" input may also be included for stronger audio signals. In all cases these ports come in the form of 1/8-inch stereo jacks.

Some desktop computers offer sound cards that plug into the motherboard. These devices are more elaborate. Some sound cards can offer as many as 12 external connections. At the rear of your computer, you may find LINE IN, MIC IN, LINE OUT, SPEAKER OUT, PCM OUT, PCM IN, JOYSTICK, FIREWIRE, S/PDIF, REAR CHANNELS or SURROUND jacks, just to name a few. For HF digital use, the important jacks are MIC or LINE IN and SPEAKER OUT.

Later in this chapter we'll discuss how to connect these sound devices to your radio, but one item needs to be briefly mentioned now: the interface. As you'll see later, the interface is yet another critical component because it is the link between the computer and the radio. The reason to mention it now is because the trend in interface technology has been to incorporate the sound device into the interface itself. These interfaces are extremely convenient because they work independently of whatever sound device you have in your computer. Just plug in their USB cables and you're good to go. It doesn't matter what kind of computer you are using; the interface will work with it. You also avoid a rat's nest of wiring between the computer and the radio. USB interfaces with built-in sound devices may cost somewhat more, but they are well worth it.

Software

When we're talking about putting together an HF digital station, a brief word about software is in order.

I say "brief" because it isn't practical to discuss software in detail within the pages of a printed book. Software evolves too rapidly for a book that has a useful lifetime measured in years. Any specific details would be obsolete almost before the book left the printing press. Instead, let's talk about software in broad strokes, beginning with operating systems.

As I mentioned earlier, most computers in amateur radio stations are running on some form of Microsoft *Windows*. Of these machines, most are using *Windows* 10 or 11.

Apple Macintosh owners are obviously quite fond of the various *MacOS* incarnations, which have proven themselves to be efficient and reliable operating systems. The universe of Mac users is growing, but amateur radio software titles for the Mac are still few in number.

The *Linux* operating system in its various versions enjoys a loyal following among amateurs who like writing their own software and tinkering at the operating system level. Like the Mac, there is ham software available, but the offerings are somewhat sparse.

The three operating systems have their advantages and disadvantages to consider from an amateur radio point of view.

Windows

Pro: Sheer variety. Most amateur radio software is written for *Windows*, so you have a rich selection of compatible software to choose from.

Con: Because of the widespread use of *Windows*, it is a favorite target for hackers. Anti-virus software is necessary, and this can significantly hamper the efficiency of *Windows* and cause other annoying issues.

MacOS

Pro: Stability and performance. Highly intuitive and easy to use. Also, hackers only rarely target *MacOS*.

Con: *MacOS* runs only on Apple Macintosh computers. It can be made to run on PCs, but it isn't an exercise for the fainthearted. Amateur radio software for *MacOS* is limited.

DigiPan **software for PSK31. This was one of the first single-mode applications for** ***Windows*****. It is still available today and will still run on modern computers.**

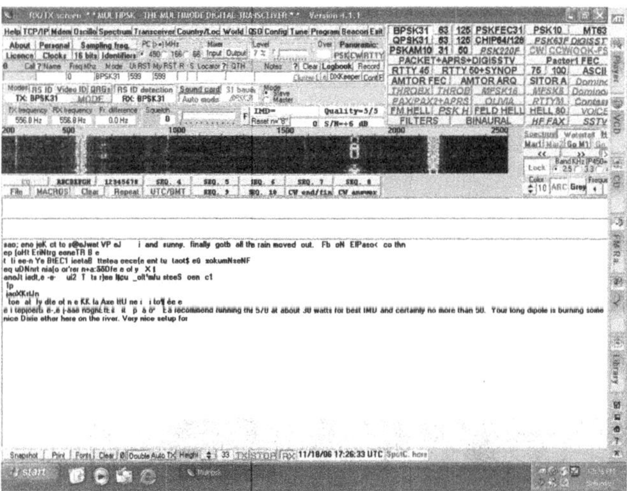

MultiPSK **for** ***Windows*** **offers many HF digital modes.**

Fldigi is a multimode application that is available for both *Windows* and *Linux*.

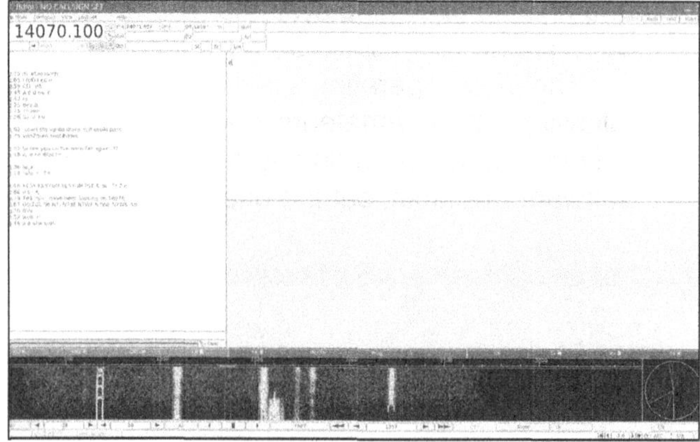

Multimode by Black Cat Systems is HF digital software for MacsOS.

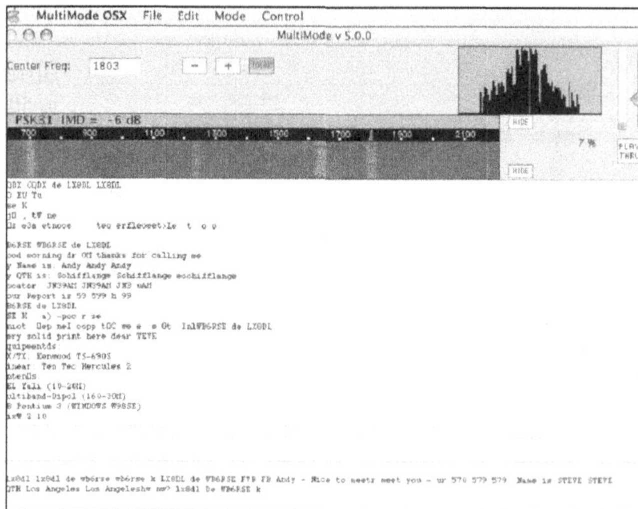

The *Digital Master 780* multimode HF digital program is part of the *Ham Radio Deluxe* package for *Windows*.

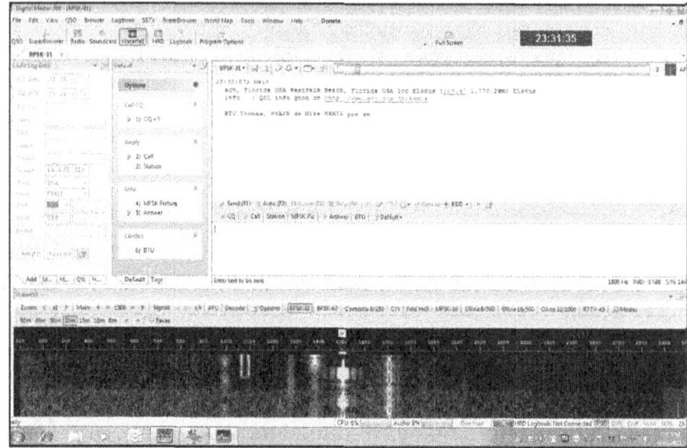

Linux

Pro: Open source and free of charge. Depending on the version, it can be quite efficient and powerful. Because there are so many versions in the field, hackers generally don't bother designing viruses for it. They prefer easier prey.

Con: Directory structures and commands may be quite different compared to *Windows.* Amateur radio software selection is limited.

Beyond the operating systems, we have specific software for HF digital operating. In the beginning programs were designed for one specific mode, such as *DigiPan* for PSK31 (**www.apkfollow.com/articles/2020/06/digipan.net.html**) or *MMTTY* for RTTY (**https://hamsoft.ca/pages/mmtty.php**). While mode-specific software still exists, the trend has been strongly in favor of multimode software that can operate many different HF digital modes.

The most popular multimode applications are:

Windows

- *MixW 4* **rigexpert.com/products/software/mixw-4/**
- *MultiPSK* **f6cte.free.fr/index_anglais.htm**
- *Fldigi* **www.w1hkj.com**
- *Ham Radio Deluxe* **www.hamradiodeluxe.com**

MacOS

- *Cocoamodem* **www.w7ay.net/site/Applications/cocoaModem/**
- *Multimode* **www.blackcatsystems.com/software/multimode.html**

Linux

- *Fldigi* **www.w1hkj.com**

Except for *Multimode, Ham Radio Deluxe,* and *MixW*, all the above are available for free downloading on the web. In later chapters we'll talk about features common to all these programs that you can use to make your operating more enjoyable.

The Interface — the Full Story

If you're like most amateurs, you already own most of the components we've discussed so far. You probably have an HF transceiver and, like 90% of US households, you have a computer in residence. The one piece of hardware that you may not own is the one item that brings these components together to create an HF digital station: the *interface*.

An interface has only one job to do: to allow the computer to

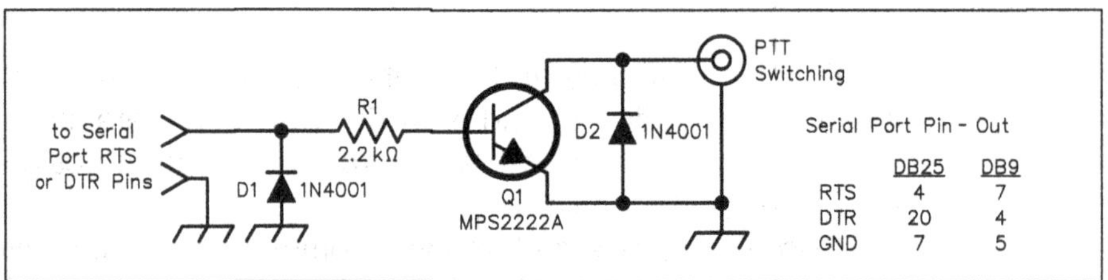

Figure 1.2 — The simplest way to key a radio using a computer is through a single-transistor circuit like this. The input connects to the serial cable coming from the computer. Either the RTS or DTR pins can be used, depending on what your software requires. The pin numbers for 25- and 9-pin plugs are shown.

toggle the radio between transmit and receive. At a very basic level, an interface can achieve this by using a signal from the computer to switch on a transistor (see **Figure 1.2**).

This transistor "conducts" and effectively brings the transceiver's *PTT* (Push to Talk) line to ground potential or close to it. When the PTT line is grounded, the transceiver switches to transmit. When the signal from the computer disappears, the transistor no longer conducts, and the PTT line is electrically elevated above ground. The result is that the transceiver returns to the receive mode.

Transceivers with Built-in Interfaces

An increasing number of HF transceivers are being manufactured with digital interfacing already built in. For those radios, the previous interface discussion is moot. All you have to do is connect a USB cable between the transceiver and your computer. Once you've configured your software to "talk" to the radio, you're all set!

Connecting an External Interface

The signal from the computer appears at a specific pin on the USB COM port. Your HF digital software generates the signal when you click your mouse on TRANSMIT or some other button with a similar label.

If an interface can be so straightforward, couldn't you just build your own? Yes, you could. Many amateurs enjoy HF digital with simple interfaces like the one shown here. In addition to the switching circuit in Figure 1.2, they connect shielded audio cables between the computer and the radio to carry the transmit and receive audio signals.

In **Figure 1.3** you'll see the three ways that most digital operators connect their computers to their transceivers. The most common means is to use an interface with a built-in sound device. However,

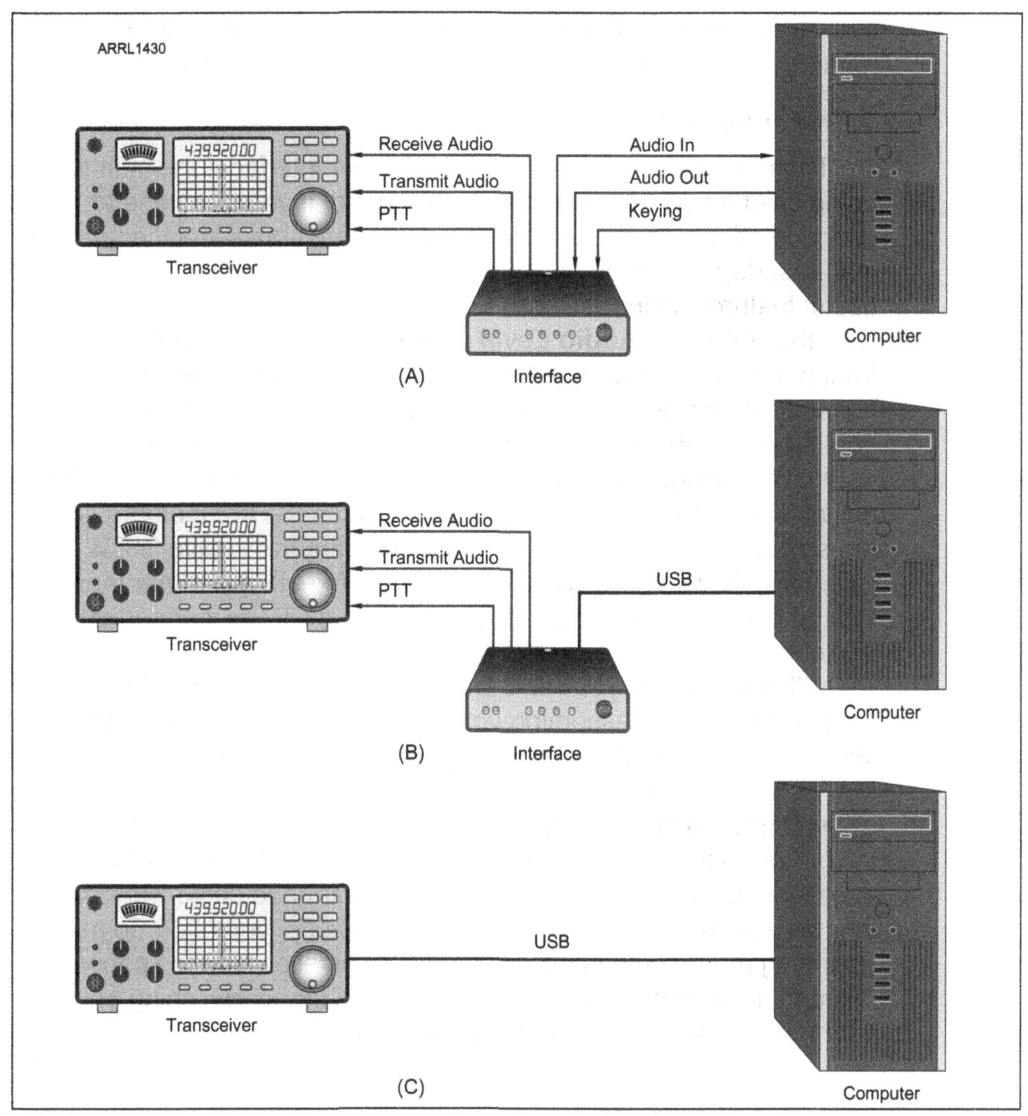

Figure 1.3 — Three of the most common configurations for interfacing your transceiver with your computer. (A) The most common setup for many years was based on an interface that took transmit and receive audio from the computer's sound card, along with transmit/receive keying from one of the computer's COM ports and passed everything to the transceiver in a way that kept the signal lines isolated. This type of interface is increasingly uncommon. (B) Many interfaces available at the time this book was written are USB devices. A single USB cable plugs into the computer. Not only is the computer's sound card not used, the interface contains its own sound device. In turn, the interface supplies transmit audio to the radio, processes receive audio from the radio and handles transmit/receive keying. (C) The future will likely see the disappearance of the interface entirely as transceivers incorporate their own computer interfacing. Today, several moderate- to higher-priced rigs have added this feature.

more transceivers are incorporating interfaces into their designs and require only a single USB cable to connect to the computer.

To Roll or Not to Roll

There are good reasons to roll your own interface, the cost savings being chief among them. On the other hand, if you purchase an interface off the shelf, you'll be able to benefit from enhanced design features, depending on how much you want to pay. The concise list of useful features includes:

• **Independent Audio Level Controls.** These are knobs on the front panel of the interface that allow you to quickly raise or lower the transmit or receive audio levels. Many amateurs prefer to manage the audio levels in this fashion compared to doing it in software.

• **CW keying.** Full-featured interfaces handle more than just HF digital. They can also use keying signals from the computer to send Morse code with a separate connection to the transceiver's CW key jack. This allows you to send CW from your keyboard rather than with a hand key, a useful feature for higher speed CW exchanges during contests.

• **FSK keying.** If you want to operate RTTY with the FSK function of your transceiver, assuming your rig offers such a function, the FSK keying feature translates keying signals from your computer into the MARK/SPACE data pulses necessary for FSK RTTY.

• **Microphone input.** If you are making HF digital connections to your radio through the microphone jack rather than the rear panel accessory port, you'll need to unplug the interface cable whenever you want to use your microphone for a voice conversation. To make operating more convenient, some interfaces allow you to keep your microphone plugged into the interface at all times, switching between your microphone or computer as necessary.

Two RigExpert TI-3000 interfaces.

• **Transceiver control.** Remember that a basic interface does not allow your computer to truly control your radio, except in the sense that it can switch your radio between transmit and receive. Deluxe interfaces include the extra circuitry needed to allow full computer control of your transceiver. Sometimes referred to as *CAT* (Computer Aided Transceiver), this is a separate function that passes all the available controls from your radio to your computer. Depending on the type of transceiver you own and the software you are using, CAT allows you to change frequency, raise and lower power levels and much more. If your transceiver has the ability to connect directly to your computer through a USB cable or Ethernet port, you don't need the CAT feature.

Manufacturers sell their own CAT interfaces, but they tend to be expensive. An interface with CAT functionality brings everything together in one affordable box.

• **Built-in Sound Device.** As we discussed earlier, several interface designs include a built-in sound device. This is particularly handy in that it liberates the sound device in your computer for other functions. You can enjoy music on your computer, for example, without having to worry that you are meddling with the sound levels you've set up for HF digital operating. In addition, an interface with a built-in sound device greatly reduces the number of cables connecting the computer and radio. The audio signals, as well as transmit/receive keying functions, are all carried over a single USB cable; there are no connections to your computer's sound ports.

The TigerTronics SignaLink USB interface also includes a built-in sound device.

• **Pre-Made Cables.** Most commercial interface manufacturers either include cables specifically wired for your radio free of charge or offer them at an additional cost. This significantly reduces the hassle of wiring your HF digital station.

At the time of this writing, a basic off-the-shelf interface costs about $80; a multi-featured deluxe interface runs as high as $400. You'll need to shop among the manufacturers to find an interface that has the features you desire at a cost you are willing to pay. Among the

The RigBlaster Advantage interface from West Mountain Radio has a built-in sound device.

more well-known popular interface manufacturers are:

MFJ: **www.mfjenterprises.com**
TigerTronics: **www.tigertronics.com**
West Mountain Radio: **www.westmountainradio.com**
RigExpert: **www.rigexpert.com**

Setting Up an Interface

If you've chosen an external interface with a built-in sound device, you'll need a set of audio and PTT cables to connect the interface to your transceiver, either at the microphone and headphone jacks, or at the accessory jack. As I've already mentioned, you can purchase this cable from the interface manufacturer or make your own. The USB cable from the interface simply plugs into your computer.

The USB connection to your computer can be a little tricky in one respect, though. When you plug the USB cable into the computer for the first time, the computer may attempt to load and run a *driver* application so it can "talk" to your interface. This driver may already exist on your computer, or you may need to download it from the manufacturer. Once the driver is loaded, the computer will recognize the interface every time you plug it in thereafter. The same advice applies if you don't have an external interface but instead own a transceiver with a built-in interface.

Regardless of whether you are using an external interface box or are fortunate to own a transceiver with interfacing built in, both connect to your computer through a USB cable and both depend on good old fashioned serial communication just as though the connection had been made through a COM port. The interface accomplishes this by creating a *virtual COM port* in your computer. In other words, it uses software to emulate the function of a COM port.

Why is it important for you to know this? The answer is that your digital software will need to be configured so that it "knows" which COM port to use for PTT keying, CAT functions, etc. That means you'll need to know this as well.

In *Windows* it is a matter of going to the **Control Panel** and hunting down the **Device Manager** icon. Once you've started Device Manager, click on the **Port** section and you'll see all your computer ports listed in order. Look for a port labeled "USB Serial Port" or "Virtual COM Port" (see **Figure 1.4**). Next to it you'll see a COM number. Write this number down because you'll need to enter it when you're setting up your HF digital program when it asks for the "serial port" or "PTT port" (**Figure 1.5**).

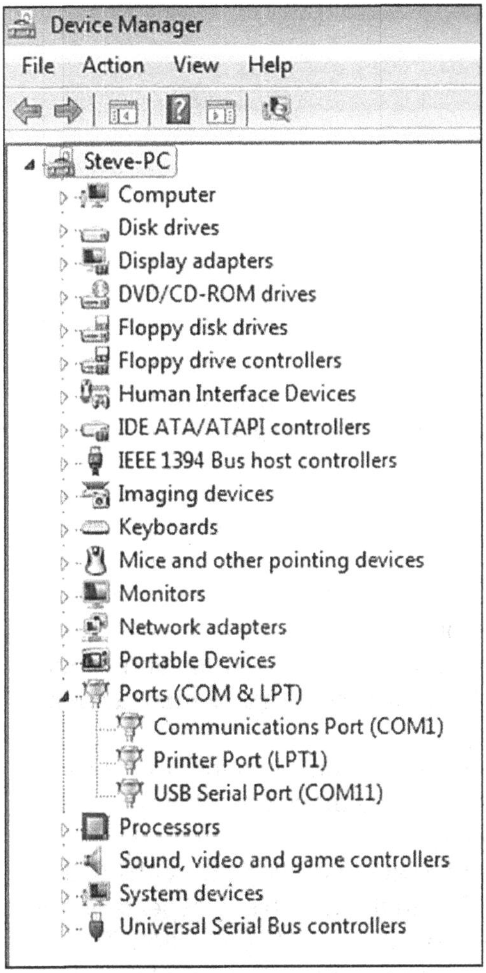

Figure 1.4 — *Windows* Device Manager showing the available ports. Note the "USB Serial Port" is labeled "COM 11."

When the computer recognizes the USB connection, it also recognizes the sound device within the interface. It will consider this device as another sound unit, just as though you had installed a second sound card inside the computer. Again, this is important to understand because when you set up your software you may need to specify which sound device the software should use. Obviously, you will need to select the sound device in your interface. Most software applications have drop-down menus that will list the available sound devices automatically. Don't expect to see your interface device listed by brand name. Instead, it may show up as "USB Sound," "USB Audio Codec" or something similar.

Managing the Audio Connections

If you own an interface with a built-in sound device, you'll need to connect the transmit and receive audio cables between

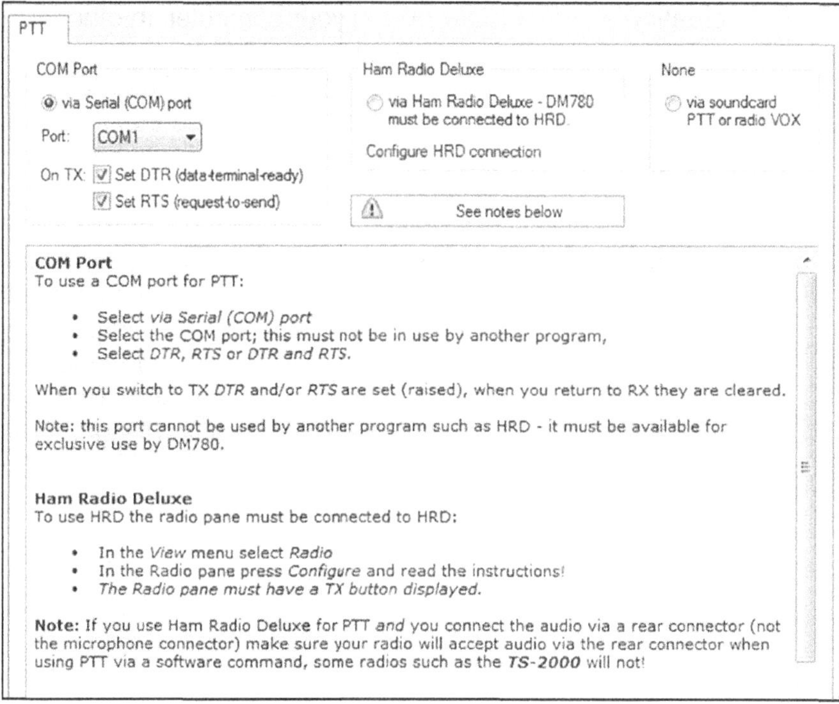

Figure 1.5 — In this configuration screen from *Ham Radio Deluxe* you'll notice that COM 1 has been selected as the transmit/receive keying port.

the interface and the transceiver, either at the microphone and headphone jacks, or at the accessory jack.

If your interface is of the simple transmit/receive switching variety, the one or both sets of audio cables may have to go all the way back to the computer.

And of course, if the interface is already built into the transceiver, you don't have to worry about audio cables at all. Everything is managed through the USB connection.

Even though you're using shielded audio and/or USB cables, there is the potential for trouble when RF is in the air. This is especially true when your station antenna is close to your operating position. The audio cables can act like antennas themselves, picking up RF and wreaking havoc on your station. I've seen some instances where the computer shut down or reset whenever the transceiver was keyed.

In other examples the RF energy mixed with the transmit audio and resulted in a horrendously distorted output.

If you suspect you have an RF interference problem in your HF digital station, you can diagnose it by reducing your output power and observing the results. If 100 W output gives you grief but 50 W is smooth as silk, you clearly have an RF interference issue.

Presumably, you've kept your audio cables are short as possible. Stringing up 20-foot-long cables between the radio and the computer is just asking for trouble.

But if your cables are of reasonable length and you still suffer interference, it is time to buy some *toroid cores*. These are circular donuts made of a powdered iron and epoxy mixture. They come in assorted sizes and are rated for suppression at various frequencies. For HF applications, Type 61 toroids are among the most effective. To suppress RF on an audio cable, wrap the cable through the toroid at least 10 times with evenly spaced turns.

With right toroid in the right place, you can greatly reduce or eliminate RF interference. For severe cases you may need to place a toroid on every cable.

You'll often see used toroids for sale at hamfest flea markets, but don't buy a used toroid unless you know the type of material it contains. When in doubt, buy toroids new from manufacturers such as Amidon at **www.amidoncorp.com**. Avoid snap-on ferrite cores. While they are certainly easy to use, they are not as effective as toroids you wind yourself.

Setting Up the Transceiver

Much of the advice that follows depends on what sort of transceiver you own and what kind of interface you are using to create your HF digital setup. When in doubt, always consult your transceiver and interface manuals.

If your interface is connecting to the radio through the transceiver accessory port, see if there is a function in the transceiver to adjust the accessory audio input and output levels. If it exists, this is a convenient way to establish "baseline" audio levels for the radio. Of course, you can also adjust audio levels at your computer and possibly at your interface (depending on the kind of interface you purchase). If it seems as though you aren't getting enough audio from the radio, or if it seems that you can't drive the radio to full output regardless of the computer or interface settings, check these transceiver settings as well.

Audio Overdrive

Speaking of driving rigs to full output, we need to discuss the danger of audio overdrive. Without question this is one of the most common issues among new HF digital operators.

Regardless of the type of digital mode you enjoy, there is an almost instinctual tendency among new operators to adjust their transmit audio levels while only watching their transceiver's RF output meter. They place their radios into the transmit mode and crank up the audio levels at their computer or interfaces until they see a satisfying 100 W RF output. At that point they assume they are finished and ready to take to the airwaves. This is known as "tuning for maximum smoke."

They could not be more mistaken!

First, most HF digital modes do not require 100 W of power to make contacts. One of the benefits of HF digital, in fact, is that you can make contacts at surprisingly low power levels.

But most importantly, adjusting for full RF output ignores the fact that you may be grossly overdriving your radio to achieve your satisfaction. The result is often a wildly distorted signal that's not only difficult to decode, but that splatters across the band, ruining every conversation in its wake.

Rather than gazing at the transceiver meter as it displays your RF

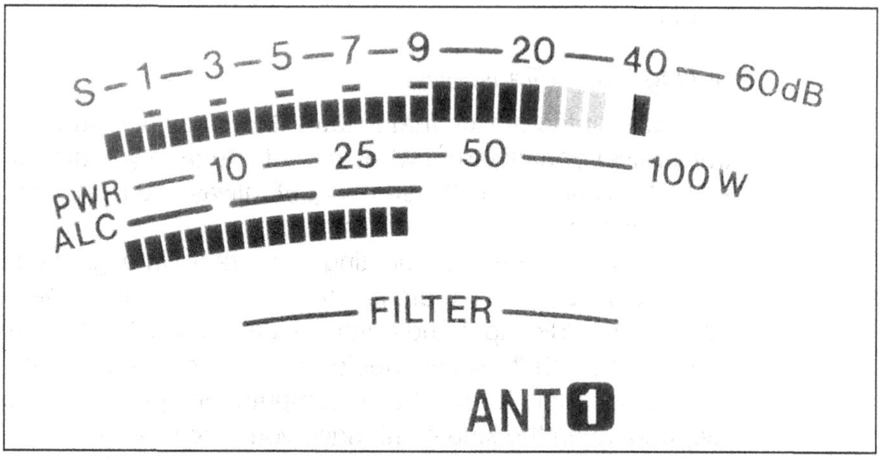

In the case of this meter, the ALC metering scale is on the bottom. As you can see, we're overdriving the radio with excessive audio, causing a substantial amount of ALC activity. Ideally, this portion of the meter should be reading zero.

In this transceiver meter, the ALC "safe zone" is indicated with a red bracket. If the needle stays within the zone, you have a decent chance of transmitting a clean signal.

output, switch the meter to monitor *ALC* (Automatic Limiting Control) instead. All transceivers display ALC activity differently. The display may simply indicate the presence and amount of transmit audio limiting taking place. Other displays may include a "safe zone." If the ALC activity remains within the safe zone, the transmit audio levels are acceptable.

When you transmit a test signal, do not increase the audio level beyond the ALC safe zone, or beyond the point where ALC activity is excessive. When you see the needle or LEDs swing hard to the right, this is a warning that you are supplying way too much audio to the radio and that the ALC circuit is trying to rein you in.

The goal is to generate the desired RF output while keeping ALC activity to a minimum (or even zero), or while keeping the ALC meter in the safe zone. It is important to keep in mind that minimal ALC activity does *not* necessarily guarantee a clean signal. It does in many instances, but your best insurance is to ask for reports whenever you are in doubt. If someone reports that you are splattering, reduce the transmit audio level until they say your signal is clean. Note this setting so that you can return to it again easily.

Computer Sounds

When you begin listening to HF digital signals, don't be surprised if you occasionally hear beep, dings, or chimes.

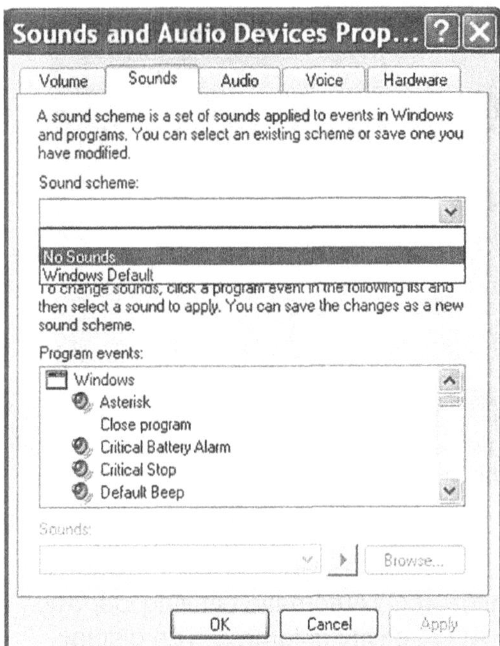

You can guard against transmitting odd *Windows* noises by turning off *Windows* sounds in Control Panel.

Most of this interference isn't deliberate. It is caused by VOX-type interfaces that key transceivers whenever they detect audio — *any audio* — from the computers. I'm willing to bet the operators aren't even aware that their computers are guilty of this obnoxious behavior.

The solution, at least in *Windows*, is simple: Turn off "*Windows* sounds" before you get on the air. You can also do this manually by opening *Windows* **Control Panel** and double clicking the **Sound and Audio Devices** icon. Click the Sound tab and under Sound Schemes select **No Sounds**. (Depending on the *Windows* version in question, the labels may differ.) Your fellow hams will thank you!

CAT Communication

If your software and interface support full transceiver control (CAT), you'll need to make sure that the data communication rates between the interface or computer and the transceiver are the same. Some clever pieces of CAT software will automatically analyze the data from the transceiver and quickly determine the data rate. For others, you'll have to enter a menu and specify the data rate.

In most CAT-capable radios you'll find a menu setting that will allow you to specify a data rate, often expressed as "baud." A rate of 9600 baud, for example, is common. You may need to access this menu to find out what setting the radio is currently using, or to change the radio's data rate to something the interface or software can handle.

When an Interface Isn't an Interface—PACTOR

Throughout this chapter we've held fast to an overarching assumption: that the mode you will choose to operate will use software that transmits and receives through a sound device. The sound device can be inside the computer, within the interface, or within the transceiver. Either way, it is at the heart of your ability to communicate.

There is one exception to this assumption: *PACTOR*.

PACTOR is a digital mode invented by two German amateurs in 1991 and it has been evolving ever since. PACTOR is capable of 100% error free communication at global distances on the HF bands.

PACTOR works its magic by sending data in small chunks or *frames*. The receiving station analyzes the data and keeps whatever has arrived without errors. It transmits a short burst back to the originating station and tells it to repeat the frame. With luck, the next retransmission of the frame will arrive with most of the data intact, or with at least enough intact data that the receiving station can mix and match between the "new" and "old" frames to produce a complete 100% error-free frame.

All this back-and-forth transmitting makes a characteristic *chirp-chirp-chirp* sound on the air. It also requires a fast-switching transceiver and a lot of computational horsepower. To this date no one has written a computer program that can do PACTOR with sound devices. The PACTOR timing parameters are too strict and computer system timing tends to be too loose.

So, the manufactures have packaged dedicated microprocessor circuitry into a stand-alone box. This box, often called a *controller*, has the speed and efficiency to meet the requirements for PACTOR. Unlike a desktop or laptop computer, a controller doesn't labor under the burden of an operating system, and it isn't required to fulfill a broad range of simultaneous tasks.

If you really want PACTOR capability, you must purchase a

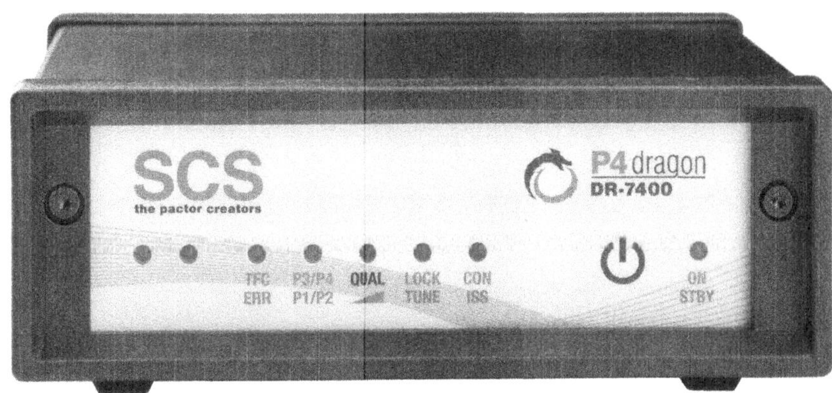

The SCS P4dragon DR-7400 is a PACTOR controller that also operates packet and RTTY.

controller. The only source is SCS Corporation at **www.scs-ptc.com**. For more information about PACTOR, see Chapter 4.

Keep in mind that to be successful with PACTOR your transceiver must be able to switch between transmit and receive in less than 20 milliseconds. Many rigs can do this, but if you are in doubt, check the *QST* magazine Product Reviews. If you are a member of the ARRL, you'll have full access to the Product Review archives online at **www.arrl.org/product-review**.

Chapter 2

RTTY

Radioteletype, better known as RTTY, is the "old man" of the HF digital modes. Despite more than seven decades of amateur use it is still a popular mode for HF digital contesting.

RTTY is quite easy to use, and every multimode digital software package listed in Chapter 1 includes RTTY in its collection. Also, there is a free, high-performance RTTY program known as *MMTTY* by Makoto Mori, JE3HHT, which we'll discuss in more detail.

Before we get started, however, let's spend some time becoming familiar with a few important RTTY terms.

Mark, Space, Shift, and Baud

RTTY uses an old digital code known as *Baudot*. Each Baudot character is composed of 5 bits. That's enough to send every letter in the English alphabet, plus the number 0 through 9 and some limited punctuation.

In amateur RTTY communication a "1" bit is represented by a 2125 Hz tone, and this is known as a mark signal. A "0" bit is represented by a 2295 Hz tone called a space signal. There is also a start pulse at the beginning of the bit string and a stop pulse at the end (see **Figure 2.1**). The rapid switching between mark and space tones gives RTTY its distinctive *deedle-deedle* sound. When you see a RTTY signal in a waterfall display, you can instantly spot the twin parallel lines representing the mark and space signals. See **Figure 2.2**.

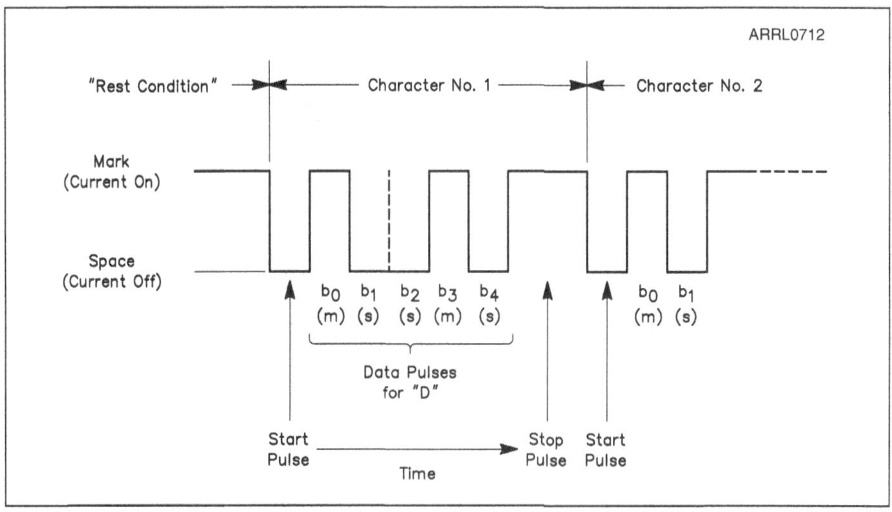

Figure 2.1 — This is a diagram of a RTTY signal as the letter "D" is sent in Baudot code. A start pulse begins the character, followed by the five bits (b0 – b4) that define it. A stop pulse signals the end of the character.

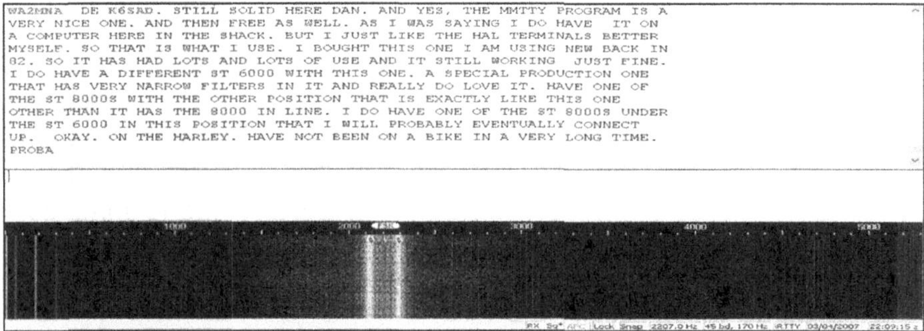

Figure 2.2 — This is a multimode program receiving a RTTY signal. You can clearly see the twin lines that represent the *mark* and *space* tones. The separation between mark and space is known as the *shift*.

As you examine Figure 2.2, you'll notice that the signal lines are separated by an "empty" area. This separation isn't random. In fact, it is highly precise. Grab your calculator and do a bit of subtraction:

2295 Hz – 2125 Hz = 170 Hz

In the answer shown above, 170 Hz is the difference or shift between the mark and space frequencies. The amateur radio RTTY standard is to use a 170 Hz shift. Maintaining proper shift is critical to RTTY communication. RTTY software that's configured for a 170

Hz shift creates narrow DSP filters that "expect" this exact frequency separation. A RTTY signal that's using a different shift will fall outside the filters and will be undecodable. FCC rules limit amateur RTTY to a 1 kHz maximum shift, but you'll rarely encounter anything other than a 170 Hz shift.

You'll also hear occasional references to a RTTY signal being "inverted." When this occurs, the mark and space tone frequencies are swapped. If the station you are communicating with is also inverted, this isn't a problem. However, it will render your text as gibberish to all the other RTTY operators who are "right side up."

When it comes to the signaling rate (or "speed," if you will), the vast majority of RTTY you'll hear on the air will be perking along at a rate of 45 baud (60 words per minute). That said, you will occasionally run into 75 baud RTTY.

On the Air with RTTY — Step by Step

If you don't already have a multimode program that offers RTTY, give *MMTTY* a try. You can download it free at **https://hamsoft.ca/pages/mmtty.php**. You will also need to download *EXTFSK*, a small COM port driver program for use with *MMTTY*, which you'll find at **https://hamsoft.ca/pages/mmtty/ext-fsk.php** (scroll down to the bottom of the page).

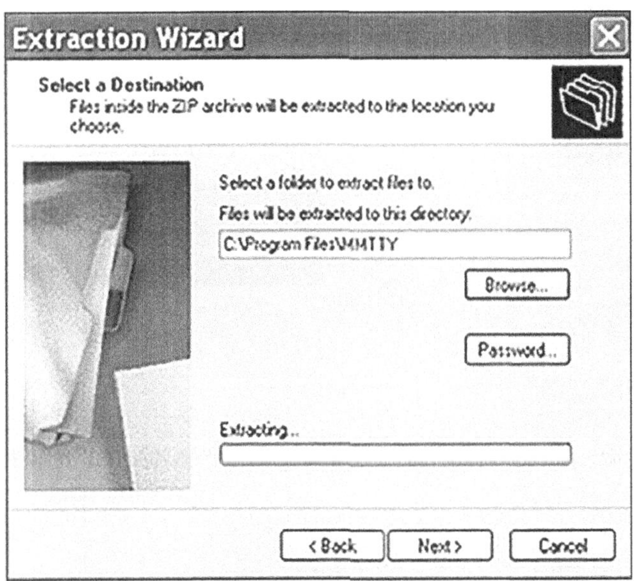

PHOTO 2A — Install *MMTTY* first, then download *EXTFSK*. Make sure to extract the files into the *MMTTY* directory on your hard drive as shown here.

Important note: Install *MMTTY* first. After installing *MMTTY*, extract the *EXTFSK* files ***to the directory where MMTTY resides***.

MMTTY is somewhat complex because it offers the ability to tweak many of its performance parameters. Even so, its *default* settings — the ones already in effect when you install the software — are perfectly adequate for our purposes. If you are using different RTTY software, you'll find that the *MMTTY* RTTY configuration is quite similar.

MMTTY Setup

Start by clicking your mouse cursor on Option(O) along the top of the *MMTTY* window. Now click on Setup MMTTY(O). A Setup window will open with seven tabs arrayed horizontally. All these tweakable settings

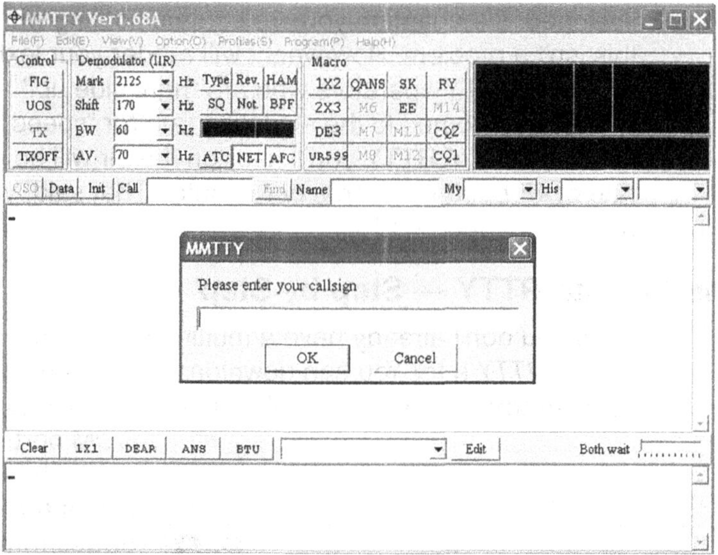

When you run *MMTTY* for the first time, you'll be asked to enter your call sign.

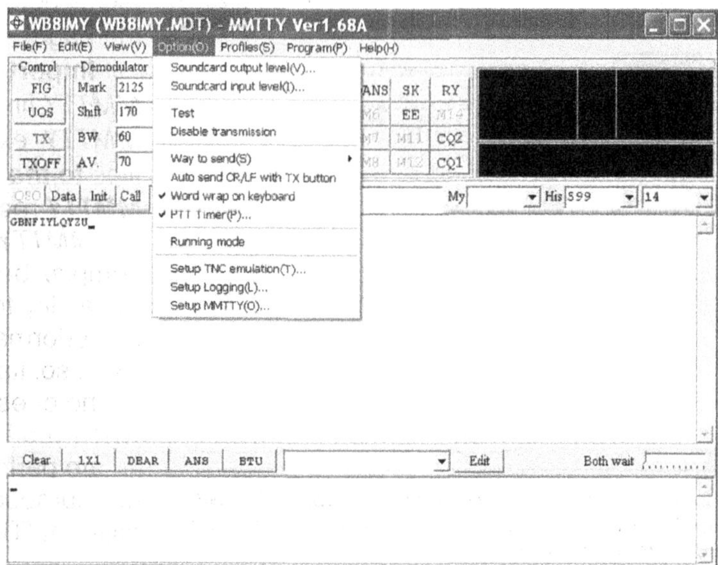

Start the setup process by selecting Setup *MMTTY* from the Option(O) menu.

may look intimidating, but we're only concerned with a few.

Click on the SoundCard tab at the far right. You'll be presented with a list of the sound devices that *Windows* has detected. Just click on the empty circles next to the sound device you are using for receive audio from your radio (audio input to the sound device) and transmit audio (audio output) going to your radio. Do ***not*** click OK — at least not yet.

Click on the TX tab. In the upper right corner of this window, you'll see an area labeled PTT & FSK. Click on the arrow in the Port window and highlight EXTFSK at the very bottom of the drop-down menu. If EXTFSK is not showing up in the Port menu that means *EXTFSK* was not properly placed in the same directory on your computer as *MMTTY.EXE*.

After EXTFSK has been selected, go to the Misc tab, and select either Sound + COM-TxD (FSK) if you are using your transceiver microphone or accessory jack as the transmit audio input, or COM-TxD (FSK) if you've set up your radio for "true" FSK keying as described in Chapter 1. If you are just dabbling in RTTY for the first time and you've already set up your station one of the other sound-device-based modes, select Sound + COM-TxD (FSK). Now click on the USB Port button to open the USB Port Option window. Select Normal in this window and click OK.

The *MMTTY* setup window. Notice the tabs along the top.

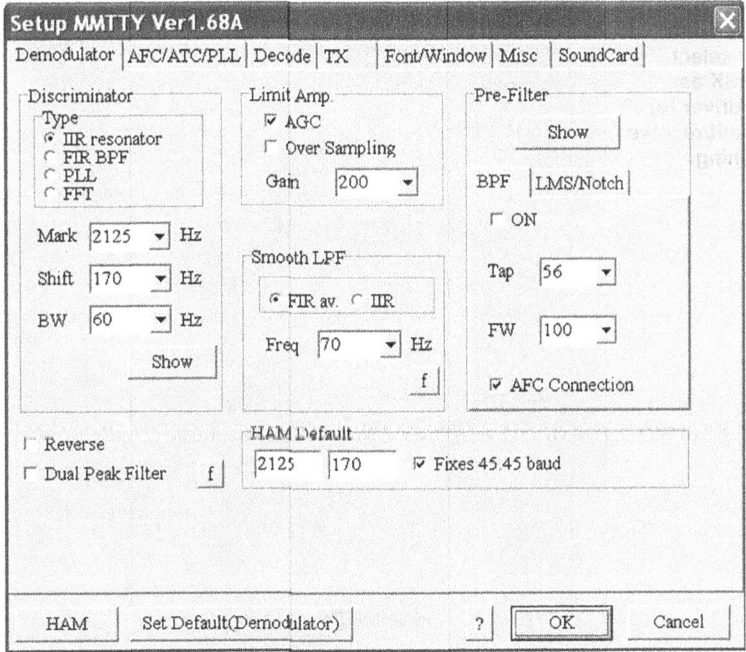

Select your sound device in this *MMTTY* window.

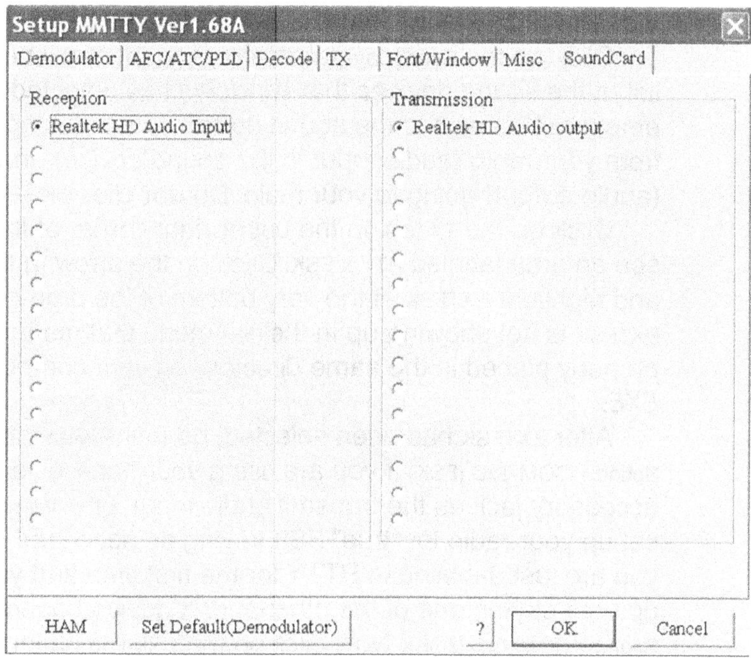

PHOTO 2F — In this menu you'll select EXTFSK as your driver for transmit/receive switching.

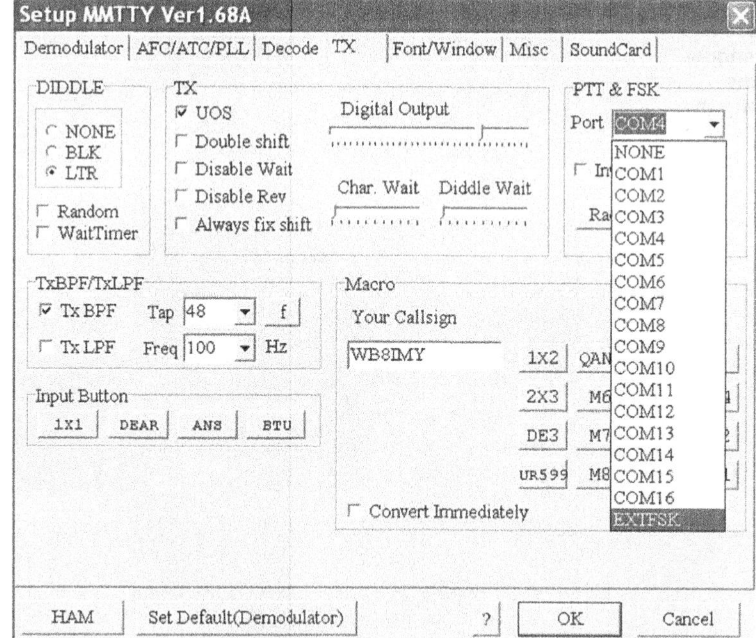

After choosing your "TX Port" option, click on the USB Port button and select "Normal" as your processing option, then click OK.

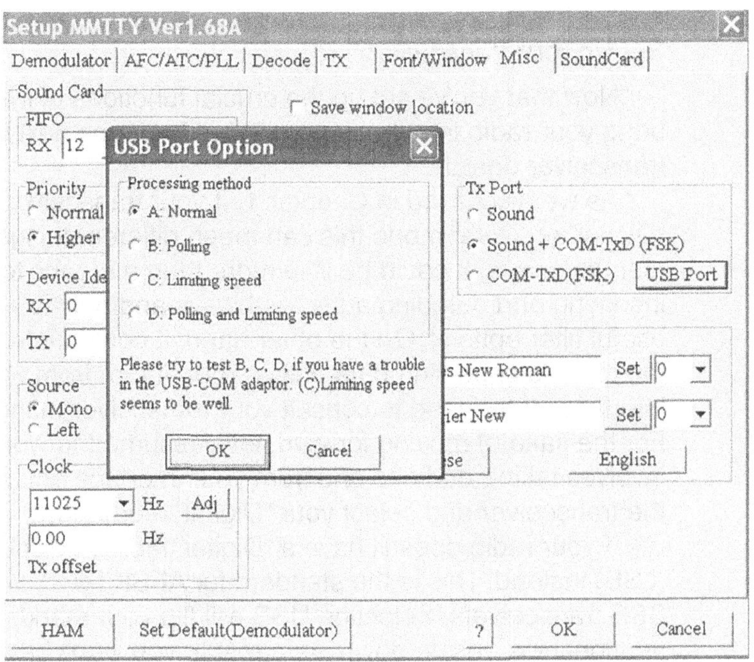

EXTFSK 1.06
Port COM1 Status:OK
FSK output: TXD, RTS, DTR
PTT output: TXD, RTS, DTR
Inv. FSK   Inv. PTT
TimeCapsMin = 1ms

The *EXTFSK* setup window. This is where you finally choose the COM port that will switch your rig between transmit and receive.

Go back to the TX tab and click the OK button. The EXTFSK window will appear separately from the MMTTY window. In this window, you can select which COM port you wish to configure. Once again, if you are using your transceiver microphone or accessory jack as the transmit audio input, select either the RTS or DTR line so that *EXTFSK* can key your transceiver's PTT (Push to Talk) line at the microphone or accessory jack.

In most cases the correct choice will probably be DTR. On the other hand, if you are using FSK for RTTY, you may need to select TxD. If you're in doubt, consult your interface manual. At worst, you may need to experiment until you find which keying line does the trick. The good news about setting up *EXTFSK* is that you only have to do it once.

The EXTFSK window will remain open while you're using *MMTTY*. If necessary, just drag it off to the side and out of the way. When you close *MMTTY*, it will close as well; it will re-appear when you open *MMTTY* again.

Set Up Your Transceiver

Now that you've set up the crucial functions of *MMTTY*, it's time to bring your radio into the picture. What happens next depends on your transceiver design.

As we discussed in Chapter 1, if your transceiver has a "RTTY," "Digital" or "Data" mode this can mean different things to different manufacturers. It could be little more than a means to select the incoming and outgoing audio pathways and possibly implement some useful filter options. On the other hand, it could be true FSK where the radio is expecting data pulses (not audio) from your computer or interface. You'll need to consult your transceiver manual to be sure. For the sake of moving forward, let's assume that your "Digital" mode involves taking audio to and from your interface and software. Turn on the transceiver and select your "Digital" mode now.

If your radio doesn't have a "Digital" mode, select Lower Sideband (LSB) instead. This is the standard for AFSK RTTY when using SSB transceivers. Selecting USB will flip your mark/space tone relationships upside down. As a result, you won't be able to decode signals, and other stations won't be able to decode your transmissions.

Take a moment to test the transmitter. Tune your radio to the middle of one of the frequency ranges in **Table 2.1** and make sure the frequency is clear. If not, pick another. Once you are in a clear spot, click your mouse cursor on the red-labeled *MMTTY* TX button. Your transceiver should switch to the transmit mode.

Table 2.1
Common RTTY Frequencies

3580 – 3600 kHz
7040 – 7100 kHz
14080 – 14099 kHz
21080 – 21100 kHz
28080 – 28100 kHz

If you are using AFSK RTTY, you need to be careful about overdriving your radio with transmit audio. While transmitting, switch your transceiver meter to ALC. If the meter is showing ALC activity, or if the ALC indicator is swinging out of the "safe" zone as described in Chapter 1, reduce your transmit audio. Depending on the type of meter your radio offers, there should be no ALC activity, or the needle should never leave the safe zone.

If you are using true FSK, you're spared from worrying about any of this. That's one of the convenient features of FSK. Your interface is sending data pulses to your transceiver and your transceiver automatically generates the proper distortion-free RF output.

This is a good time to remind you about RF output power. Remember that RTTY is a 100% duty cycle mode. If your rig is designed to handle 100% duty cycle signals, you're safe. Otherwise, you'll need to reduce your RF output by as much as 50%. Check your transceiver manual.

Click the TXOFF button to return to receive.

Tuning and Decoding RTTY Signals

Every RTTY decoder, whether it works in software or hardware, incorporates a set of mark and space audio filters. You can think of these filters as dual windows that only open for tones that are at the correct mark and space frequencies and separated by the proper shift. The mark and space filtering circuitry detects and decodes the tones into digital 1s and 0s, which is exactly what your computer needs to provide text on your screen. The more sensitive and selective your mark/space detectors, the better your RTTY performance, especially as it involves your ability to copy weak signals through interference.

It's easy to understand why tuning a RTTY signal (and just about any other data signal) is so critical—and why a good tuning indicator is one of your best HF digital tools. Whenever you stumble upon a RTTY signal, you must quickly tune your receiver until its mark and space tones fall within the "skirts" of the filters and are detected. It's possible

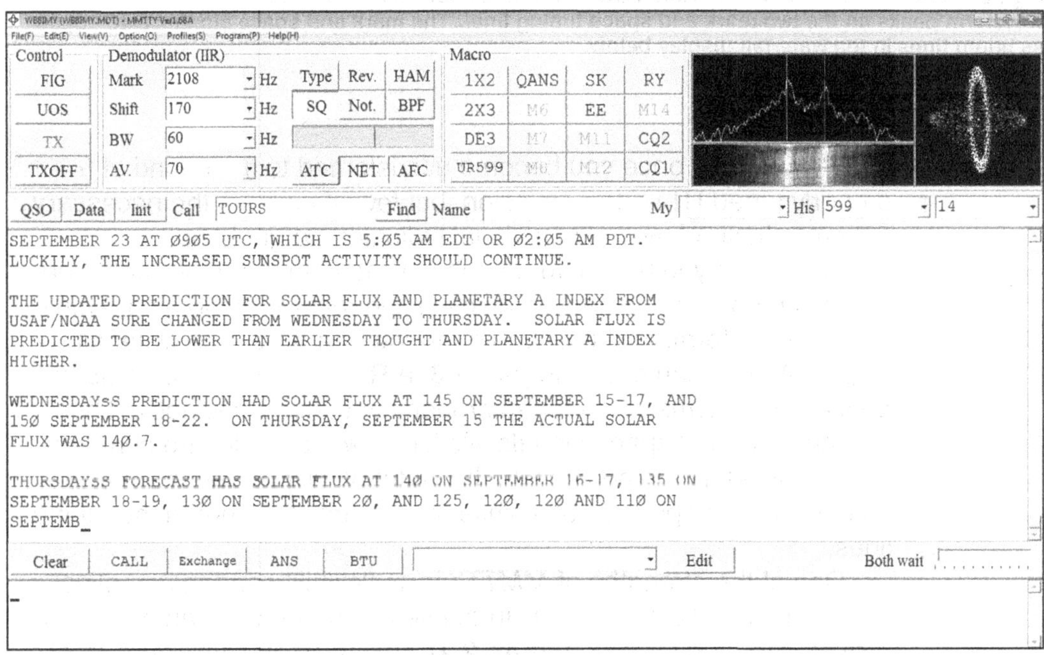

Receiving a RTTY bulletin from ARRL Headquarters station W1AW.

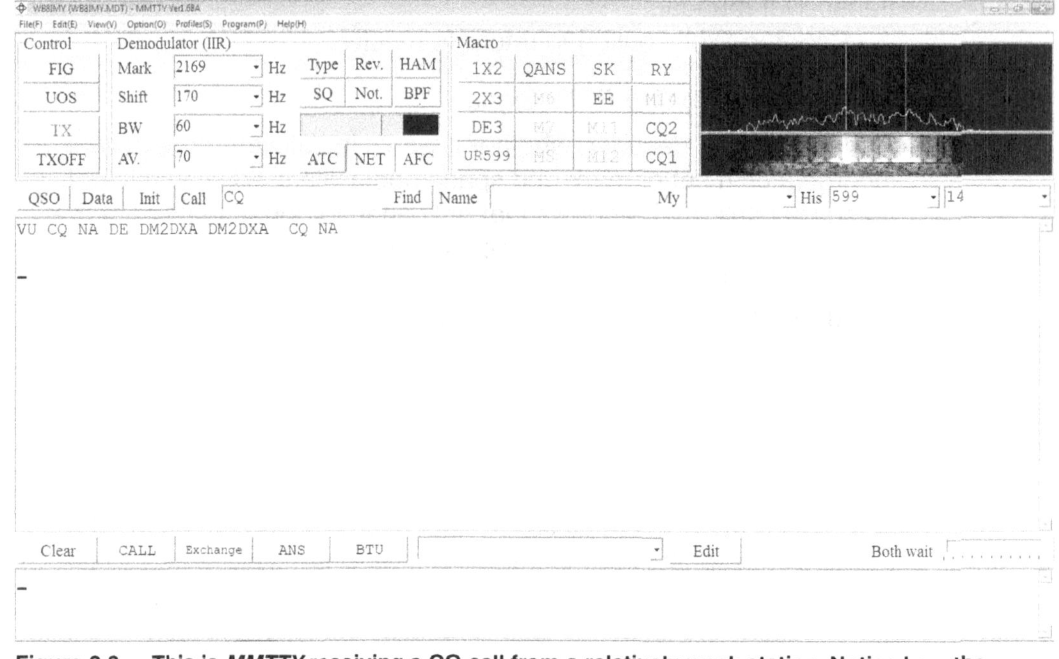

Figure 2.3 — This is *MMTTY* receiving a CQ call from a relatively weak station. Notice how the waveform peaks on the two mark and space tuning lines. The mark and space signals also appear as bright lines in the waterfall display below.

to do this by ear once you become accustomed to the sound of a properly tuned RTTY transmission, but few of us have the necessary patience. Instead, we rely on visual indicators to guide us.

In *MMTTY* you have two default indicators in the upper right corner of the screen. The top section displays the receive audio as a rapidly varying waveform. Just below this window is a grayscale waterfall display. As you can see in **Figure 2.3**, a RTTY signal appears as two peaked waveforms when tuned properly in the upper window. These are the mark and space signals. Notice how the peaks are centered on the two white tuning lines. In the waterfall display the mark and space signals appear as two white traces directly beneath the tuning lines.

One of the strengths of *MMTTY* is its flexibility and that applies to the tuning indicator as well. In the View (V) menu you can choose to activate the XY Scope (see **Figure 2.4**). This is a software variation of what RTTY old timers used to call the "crossed bananas" display. You

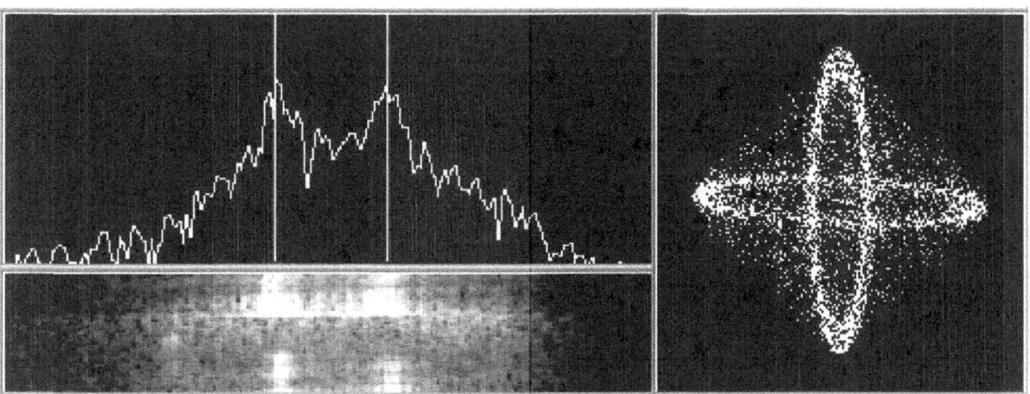

Figure 2.4 — The *MMTTY* tuning indicator with the XY Scope display activated. In addition to the sharp mark and space peaks in the spectral display, the ellipses are crossed in the scope window.

tune the RTTY signal until the ellipses cross.

Unless there is a contest or a DX pileup going on, finding RTTY on the air may require some patience. Take another look at the frequencies in **Table 2.1**. This is where you are most likely to encounter RTTY on any given day. Twenty meters is the most popular band. Another option is to monitor a W1AW RTTY transmission from ARRL Headquarters. You'll find the W1AW bulletin schedule online at **www.arrl.org/digital-transmissions**. Note that RTTY transmissions are referred to as "Baudot" in the W1AW schedule.

Once you've found a signal, tune it in carefully and watch the text march across the large receive window. *MMTTY* is sensitive software, and it tends to respond to random noise by printing equally random characters in the receive window. If this annoys you, activate the software squelch function by clicking your mouse cursor on the SQ button.

As you watch the signal, notice how the conversation flows just like a voice or CW rag chew.

KF6I DE WB8IMY . . . YES, I HEARD FROM SAM JUST YESTERDAY. HE SAID THAT HIS TOWER PROJECT WAS ALMOST FINISHED. KF6I DE WB8IMY K

If you want to call CQ, the procedure is simple. *MMTTY* has two "canned" CQ messages (known as *macros*) that you can send by clicking on either the CQ1 or CQ2 buttons. You can right click on these buttons can change the macros to your liking.

Alternatively, you can manually type your CQ in the bottom transmit buffer window. When you click the TX button, *MMTTY* will send everything in the transmit buffer.

For now, let's try the CQ1 button. Click on it and you'll hear the delightful chatter of RTTY and see the following on your screen (with your call sign instead of mine):

CQ CQ CQ DE WB8IMY WB8IMY WB8IMY PSE K

Notice how the CQ is short and to the point. Did you also notice that it repeated the call sign several times? *Remember that RTTY lacks error detection.* If you want to make certain that the other station copied what you sent, it helps to repeat it. You'll see this often in contest exchanges. For example:

N6ATQ DE N1RL . . . UR 549 549. STATE IS CT CT. DE N1RL K

On the other hand, if you know that the other station is copying you well, there is no need to repeat information.

An option to investigate in your RTTY software is Unshift On Space (UOS). There is a benefit to using UOS when receiving. Two special characters, LTRS and FIGS, are used by the RTTY code to indicate to the computer or processor whether the characters that follow will letters or figures (numbers and punctuation). If you enable UOS, your software will exit the FIGS shift as soon as there is a space, so your display does not print out garbage if the software misses the LTRS shift character because of noise or interference. In *MMTTY* UOS is the default.

MMTTY has an Automatic Frequency Control (AFC) function that will attempt to track the received signal as it drifts due to propagation conditions or other factors. You can disable this by clicking on the AFC button. The adjacent NET button has a similar effect on your transmitted signal.

AFC and NET are both handy functions, but they can cause problems when not used correctly. If you are enjoying a casual RTTY conversation on an uncrowded band, you can leave AFC and NET on. But if the band is crowded — during a RTTY contest, for example — you may be better off turning both functions off. Many RTTY operators run very tight filters in crowded band conditions, some as narrow as 250 Hz. If the AFC and NET are on, *MMTTY* could easily shift your signal outside the range of the other station's filters. Similarly, you could be calling CQ with a narrow IF filter and wondering why no one replies!

Chasing RTTY DX

Like any other form of DXing, the quest for RTTY DX demands patience and skill. When a DXpedition is on the air with RTTY from a rare DXCC entity, your signal will be in competition with thousands of other HF digital operators who want to work the station as badly as you. Sometimes pure luck is the winning factor, but there are several tricks of the trade that you can use to tweak the odds in your favor.

Don't Call... Yet

Let's say that you're tuning through the HF digital subbands one day and you stumble across a screaming mass of RTTY signals. On your computer screen you see that everyone seems to be frantically calling a DX station. This is known as a *pileup*.

You can't hear the DX station that has everyone so excited, but what the heck, you'll activate your transceiver and throw your call sign into the fray, right? *Wrong!*

Never transmit even a microwatt of RF until you can copy the DX station. Tossing your call sign in blindly is pointless and will only add to the pandemonium. Instead, take a deep breath and wait. When the calls subside, can you see text from the DX station on your screen? If not, the station is probably too weak for you to work (don't even bother), or the DX may be working "split." More about that in a moment.

If you can copy the DX station, watch the exchange carefully. Is the station calling for certain stations only? In other words, is the DX sending instructions such as "North America only"? Calling in direct violation of the DX station's instructions is an effective way to get yourself blacklisted. Does the DX just want signal reports, or are they in the mood for brief chats? Most DX stations simply want "599" and possibly your location—period. Don't give them more than they are asking for. (A DX RTTY station on a rare island doesn't care what kind weather you are experiencing now.)

Working the Split

When DX RTTY pileups threaten to spin out of control, many DX operators will resort to working *split*. In this case, "split" means split frequency. The DX station will transmit on one frequency while listening for calls on another frequency (or range of frequencies).

Good DX operators will announce the fact that they are working split with almost every exchange. That's why it is so important to listen to a pileup before you throw yourself into the middle. If you tune into a pileup and cannot hear the DX station, tune below the pileup, and see if you copy the DX station there. If their signal is strong enough, they shouldn't be hard to find if they are working split. Their signal will seem to be by itself, answering calls that you cannot hear. This is a major clue that a split operation is taking place. Watch for copy such as:

CQ DX DE FOØAAA, UP 10 (Translation: DX is listening up 10 kHz.)

CQ DX DE FOØAAA, 14085-14090 (Translation: DX is listening between 14085 and 14090 kHz.)

Whatever you do, *never call a split DX operation on the station's transmitting frequency*. Your screen will quickly fill with rude comments from others who are listening. Instead, put your transceiver into the split-frequency mode (better drag out your user manual if you're unsure how to do this). Set your radio to receive on the DX station's transmitting frequency and transmit on their listening frequency. If the DX station is listening through a range of frequencies, you'll need to select a spot where you think you'll be heard. Change your transmit frequency if this particular "fishing spot" doesn't seem to be working.

Short and Sweet

When you've done your listening homework and you're ready for battle, by all means fire at will. Wait until the DX station finishes an exchange. They'll signal that they're ready for another call by sending "QRZ?" or something similar. When it's time to transmit, make it short and to the point, like this:

WB8IMY WB8IMY WB8IMY K

Listen again. Has the DX responded to anyone yet? If the answer is "no" and other stations are still calling, the DX is probably trying to sort out the alphabet soup of confusion on their screen. Try again.

WB8IMY WB8IMY WB8IMY KK

You might make it through at just the right moment when other signals subside briefly, or when the ionosphere gives you an unexpected boost. But if no one is calling, or if you hear the DX station calling someone else, *stop*. You lost this round, so give the lucky winner their chance to be heard. Your next opportunity will be coming up shortly.

Did the DX Station Call You?

Watch your screen carefully. If the DX station only copied a fragment of your call sign, the DX might send something like:

IMY IMY AGAIN??

In this instance DX copied only the last three letters of my call sign ("IMY"). Fading and flutter can make RTTY signals difficult to copy clearly. The best thing to do is reply right away, sending your call sign three times just like before.

If luck is on your side, you'll see:

WB8IMY DE FO∅AAA . . . TNX. 599. QSL? K

And with excited fingers you answer:

QSL. UR 599. TNX AND 73 DE WB8IMY K

RTTY Contesting

Nothing tests your equipment and operating skills like a contest. See the list of major annual RTTY contests in **Table 2.2**.

A contest really shakes the bugs out of your station. If you have shortcomings in your antenna system, you're going to discover them very quickly. If your transceiver can't seem to handle the crunch of dozens of signals in proximity, it will become painfully apparent within

Table 2.2
Major RTTY Contests Throughout the Year

For details, see the ARRL website at www.arrl.org/contest-calendar or the WA7BNM contest calendar at **www.contestcalendar.com**.

New Year's Day	SARTG New Year RTTY Contest
First weekend in January	ARRL RTTY Roundup
Third weekend in January	BARTG RTTY Sprint
Second weekend in February	CQ World Wide WPX RTTY Contest
Fourth weekend in February	North American QSO Party
Second weekend in March	North American RTTY Sprint
Third weekend in March	BARTG HF RTTY Contest
First weekend in April	EA RTTY Contest
Second weekend in May	A. Volta RTTY DX Contest
Third weekend in July	North America QSO Party
Third weekend in August	SARTG RTTY Contest
Last weekend in September	CQ WW RTTY DX Contest
Second weekend in October	BARTG RTTY SPRINT
Third weekend in October	JARTS World Wide RTTY Contest
Second weekend in November	Worked All Europe DX Contest
Third weekend in December	OK DX RTTY Contest

minutes. If you need better logging or contest software, the first hour of a contest will provide a powerful motivation to upgrade.

Some hams shun contesting because they assume they don't have the time or hardware necessary to win—and they are probably right. But winning is *not* the objective for most contesters. You enter a contest to do the best you can, to push yourself and your station to whatever limits you wish. The satisfaction at the end of a contest comes from the knowledge that you were part of the glorious frenzy, and that you gave it your best shot.

Contesting also has a practical benefit. If you're an award chaser, you can work many desirable stations during an active contest. During the ARRL RTTY Roundup, for example, some hams have worked enough international stations to earn a RTTY DXCC. Jump into the North American RTTY QSO Party and you stand a good chance of earning your Worked All States in a single weekend.

Always remember that contesting is ultimately about *enjoyment*. The thrill of the contest chase gets your heart pumping. The little triumphs, like working that distant station even though they were deep in the noise, will bring smiles to your face.

Digital Contesting Tips for the "Little Pistol"

If you are like most amateurs, your station falls into the Little Pistol category. The Little Pistol station usually includes a low- to medium-priced 100-W transceiver feeding an omnidirectional antenna such as a dipole, vertical, end-fed wire, and so on. Even if you are blessed with a beam antenna on a tower or rooftop, in all likelihood you are still a Little Pistol.

At the opposite end of the spectrum are the Big Gun stations. They have multiple beams on tall towers, phased array antennas on the lower bands, incredibly long Beverage receiving antennas and more. They own high-end (read: expensive) transceivers that boast razor-sharp selectivity, and they usually couple them to 1.5 kW linear amplifiers.

So how does a Little Pistol go about beating a Big Gun in a digital contest? With rare exceptions, the Little Pistol can't. Banging heads with a Big Gun station is usually an exercise in futility. The Big Gun is going to hear more stations, and more stations are going to hear it. Focus, instead, on using your wits and equipment to get the best score possible. You probably won't beat a Big Gun, but you could make him sweat!

Check Your Antenna

How long has it been since you've inspected your antenna? Is everything clean and tight? And have you considered trying a different antenna? Contests are terrific testing environments for antenna designs. For example, if you've been using a dipole for a while, why not try a loop? Just set up the antenna temporarily for the duration of the contest and see what happens.

Check Your Equipment

If you are having problems with RF getting into your computer or sound card, fix these glitches *before* you find yourself in a contest. Also, consider spending $140 or so to install a 500-Hz IF filter in your radio if it doesn't have one already, or if it lacks built-in IF DSP filtering. Finally, check your computer and software. Make sure you understand the program completely. Set up any "canned" messages/responses and test them before the contest begins.

Understand Propagation

If you are participating in a DX contest, is it a good idea to prowl the 40-meter band in the middle of the day? No, it isn't. Forty meters is only good for DX contacts after sundown. If you are hunting Europeans or Africans on 10 or 15 meters, when is the best time to look for them? The answer is late morning to early afternoon.

In other words, let propagation conditions guide your contest strategy. Your goal is to squeeze the most out of every band at the right time of day. For example, you might start the ARRL RTTY Roundup on 10 meters and then bounce between 10, 15, and 20 meters until sunset. After sundown, concentrate on 20 and 40 meters. In the late night and the wee hours of the morning, it is best to try 40 and 80 meters.

In some contests you can enter as a single-band station. Which band should you choose? That depends on the contest and your equipment. If you have a 15-meter beam and you are entering a DX RTTY contest, it makes sense to concentrate on 15 meters, even though you'll probably find yourself with nothing to do in the late evening after the band shuts down. Twenty meters makes sense as the ultimate "around the clock" band, but it is often very crowded during contests and populated with Big Gun stations. In some instances, you may find it more "profitable" to concentrate your efforts on another band without so many signals.

Hunting and Pouncing vs. "Running"

The commonsense rule of thumb is that a Little Pistol station should only hunt and pounce. That means that you patrol the bands, watching for "CQ CONTEST" on your monitor, and pouncing on any signals you find. The Big Guns, on the other hand, often set up shop on clear frequencies and start blasting CQs. If conditions are favorable, they'll be hauling in contacts like a commercial fishing trawler! This is known as *running*.

In many cases Little Pistols are probably wasting valuable time by attempting to run. There are situations, however, where running *does* make sense. If you've pounced on every signal you can find on a particular band, try sending several CQs yourself to catch some of the other pouncers. If you send five CQs in a row and no one responds, don't bother to continue. Move to another band and resume pouncing.

Seek the Multipliers

Every contest has multipliers. These are US states, DXCC entities, ARRL sections, grid squares and so on, depending on the rules of the contest. A multiplier is valuable because it multiplies your total score.

Let's say that DXCC countries are multipliers for our hypothetical contest. You've amassed a total of 200 points so far, and in doing so you made contacts with 50 different DXCC entities.

$200 \times 50 = 10{,}000$ points

Those 50 multipliers made a dramatic difference in your score! Imagine what the score would have been if you had only worked 10 multipliers?

If given a choice between chasing a station that won't give me a new multiplier and pursuing one that *will*, I'll spend much more time trying to bag the new multiplier.

Contest Software

No one says you must use software to keep track of your contest contacts, but it certainly makes life easier! One of the fundamental elements of any contest program is the ability to check for duplicate contacts or dupes. Working the same station that you just worked an hour ago is not only embarrassing, it is a waste of time. The better contest programs feature automatic dupe checking. When you enter the call sign in the logging window, the software instantly checks your log and warns you if the contact qualifies as a dupe. The more sophisticated programs "know" the rules of all the popular digital

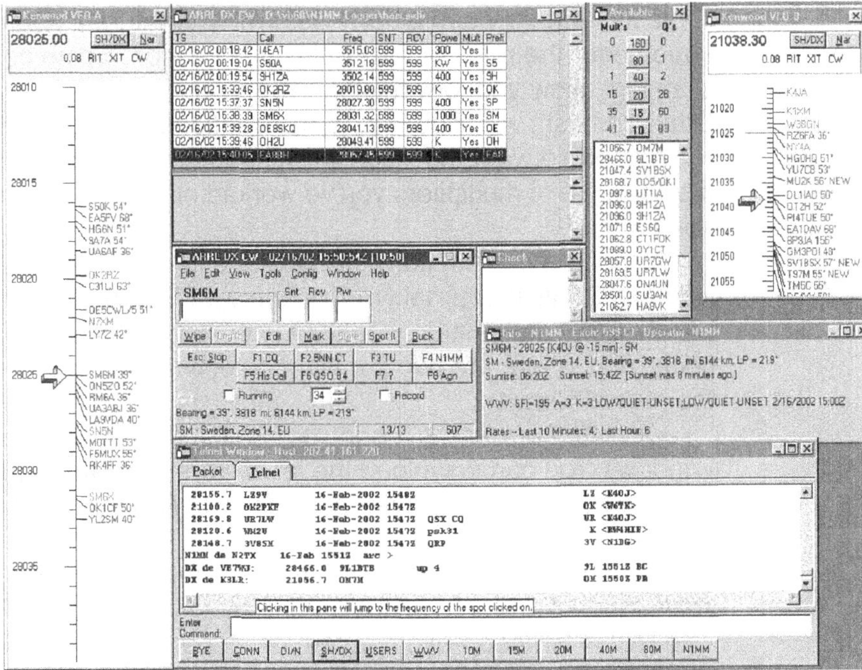

N1MM contest logging software.

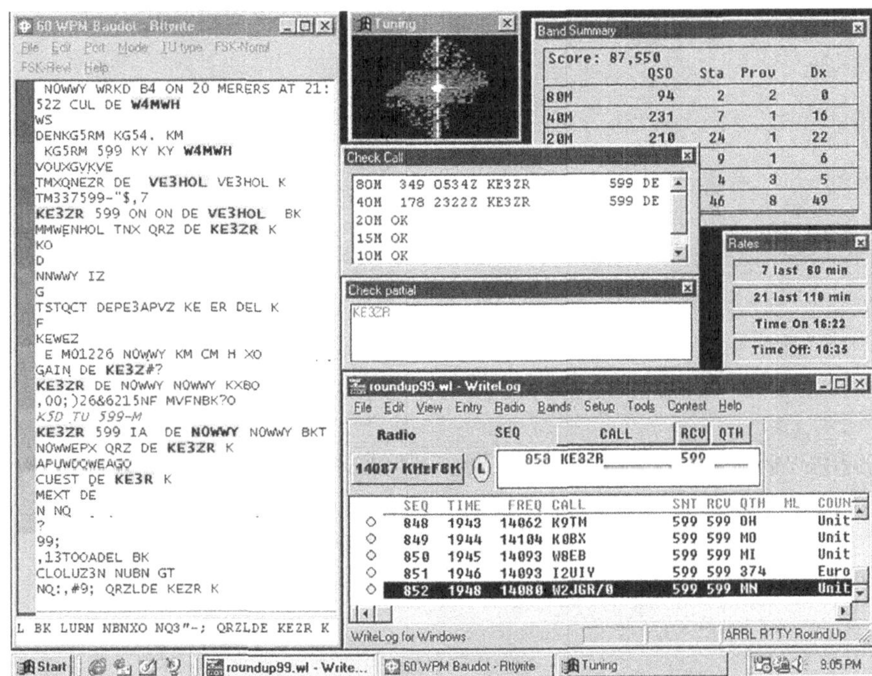

WriteLog in action.

contests, and they can quickly determine whether a contact is truly a dupe under the rules of the contest in question. Some contests, for example, allow you to workstations only once, regardless of the band. Other contests will allow you to workstations once per band.

A good software package will also help you track multipliers. It will display a list of multipliers you've worked or show the ones you still need to find.

Any of the contest software packages advertised in *QST* and the National Contest Journal will function well. To minimize headaches, however, my suggestion is to stick with software written specifically for contesting. There are two popular contesting software packages on the market today:

• *N1MM Logger+* is a popular free contesting application for Windows. You can download the latest version at **https://n1mmwp.hamdocs.com**. *N1MM Logger+* supports *RTTY* contesting with *MMTTY*.

• *WriteLog* by Ron Stailey, K5DJ, is a full-featured Windows-based software package with an interesting twist — a built-in *RTTY* program. Like *RTTY*, *WriteLog* offers automatic dupe checking and flagging, multiplier displays, radio control, canned messages, and more. See the *WriteLog* site on the Web at **www.writelog.com**.

Chapter 3

FT8 and FT4

When you hear someone referring to the "*WSJT* modes," they are talking about the digital operating modes included in the *WSJT* software suite created by Dr. Joe Taylor, K1JT. *WSJT* — now known as *WSJT-X* — offers several different modes, each with its own strengths for a given application.

WSJT had its beginnings as a collection of software for moonbounce communication. As a Nobel Prize-winning scientist who studies pulsars and other distant astronomical objects, Joe has a keen interest in weak signals, the kind hams encounter when reflecting signals from the lunar surface. Joe's software exploits the power of modern desktop and laptop computers to separate weak signals from noise and decode the information they contain. With just a sound card or sound chipset and a transceiver interface, *WSJT* made it possible for hams with modest stations to enjoy VHF meteor scatter communication and even moonbounce. Prior to Joe's invention, hams needed to make large investments in antennas, transceivers, and amplifiers to pursue these activities. The advent of *WSJT* opened an exciting world of operating for many who could not otherwise afford it.

But it wasn't long before someone wondered what would happen if one of the *WSJT* modes, known as JT65, was used on the HF bands. Digital communication on HF isn't as challenging as getting a signal to the Moon and back, so it made sense that there would be plenty of "performance margin" to provide fascinating results. To no one's surprise, this turned out to be true. Using a variant of JT65 known as JT65A, even a few watts of JT65-modulated RF to a wire dipole antenna resulted in transcontinental and even global communication.

JT65 became a popular HF digital operating mode in short order and, by 2012, it even eclipsed PSK31. Then, in 2017, a version of *WSJT-X* was released that included a new mode: FT8.

For reasons that will become clear in this chapter, FT8 took off like a proverbial rocket. It offered many of the benefits of JT65 but was much faster and semiautomated. By 2022, FT8 had become the most widely used digital mode on the HF bands and it was joined by its cousin, FT4.

A Word About Software

By 2018, *WSJT-X* had achieved a position of dominance in weak-signal HF digital world. It is easy to understand why this was so. With *WSJT-X*, you have all the popular weak-signal HF digital modes in a single application. *WSJT-X* contains not only FT8 and FT4, but also the widely used WSPR beacon mode, JT9, JT65, MSK144 for meteor scatter, and several others.

With this in mind, we'll devote this chapter strictly to a discussion of *WSJT-X* because that is the software package most readers will be using. And rather than exploring every operating mode *WSJT-X* has to offer, we will concentrate on FT8 and FT4.

One disadvantage of discussing software in a printed book is that the software tends to change much faster than the book. *WSJT-X* is no exception, especially as Dr. Taylor and his team continue to explore new operating modes and perfect existing ones. However, it is safe to say that *WSJT-X* will not change so dramatically that it renders this chapter obsolete before it can be revised in a subsequent edition. You may need to improvise and adapt the instructions to a future *WSJT-X*, but much of what you will find in this chapter will still apply.

Getting Started with FT8

FT8 is not a "conversational" mode. In fact, there isn't a conversation taking place at all. Instead, all you are doing is making contact and exchanging signal reports and grid square locations. So, what is the attraction of FT8 and FT4?

• Both modes provide "honest," objective signal reports. The software automatically makes a signal-to-noise measurement and generates a report for the other operator. Signal strength is expressed in dB (Decibels). For example, a -25 dB signal is quite weak while a +4 dB signal is strong. These reports give you a true sense of how well your signal is being received. This is critical when testing antennas or

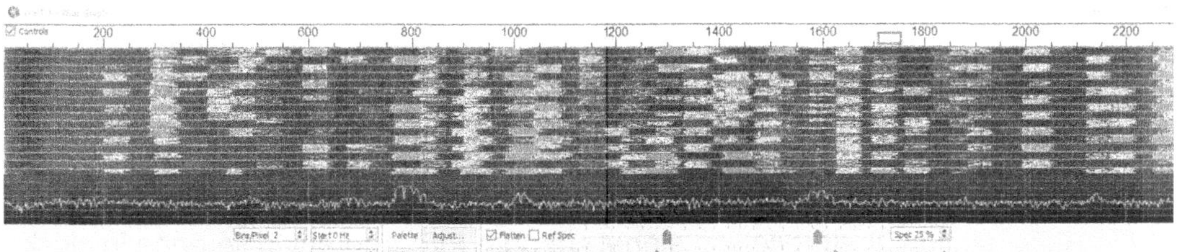

Figure 3.1 — The *WSJT-X* waterfall/spectrum window as recorded with the VFO set for 14.074 MHz (upper sideband). The numbers above the waterfall represent the audio frequencies in Hz.

studying propagation.

• Their performance is such that even a minimal station, such as someone transmitting 5 W to an attic antenna, can make large numbers of contacts, including contacts with stations in other countries.

• Their exchanges provide all the information needed for valid contacts that can be applied to awards such as the ARRL DX Century Club.

And for FT4, an additional attraction is speed. It is possible to complete an FT4 contact in little more than 36 seconds.

To get started with FT8 or FT4, you must first download and install *WSJT-X*. While most hams use the Microsoft *Windows* operating system, you will also find versions of *WSJT-X* for *Mac OS* and *Linux* at **https://physics.princeton.edu/pulsar/k1jt/wsjtx.html**.

When you start *WSJT-X*, you'll see two windows: a main window with menus for the settings, options, and the information being decoded, and a separate window for the combined waterfall and spectrum display. Each window can be moved on your monitor and resized to suit your preference and the monitor's resolution.

Figure 3.1 shows the *WSJT-X* waterfall/spectrum window as recorded with the VFO set for 14.074 MHz (upper sideband). The numbers above the waterfall represent the audio frequencies. The vertical axis in the *waterfall* represents time in 15 second intervals. The vertical axis in the *spectrum* window represents the signal strength (dB). There are options to adjust the bins/pixel, start frequency, and averaging pattern, but you can leave these at their default settings for now.

The signal strength of each FT8 signal can be estimated by its peak amplitude relative to the average noise floor. Two FT8 signals will often overlap in frequency, but the software does an excellent job separating the individual messages.

Figure 3.2 shows activity on 20 meters. The main window's decoded transmissions are in 15 second time intervals. You can see the band activity at the lower left showing all stations copied, listed in order of time and audio frequency.

You can read the call signs, the relative audio frequency of each station, their received signal strength (dB) and any difference in clock accuracy (DT) between the station you are receiving and your own computer's clock (seconds).

Each message (shown in the Generate Std Msgs box at the lower right-hand portion of the screen) is sent in the next available 15 second window after you receive a response from the station you are working.

At the far lower left is a vertical scale, 0 to 90 showing the total strength of *all* received audio signals (like an S meter on your rig), which at this moment reads 65 dB. You don't want too much, or too little received audio.

The *WSJT-X* MONITOR button must be engaged to receive. The Enable Tx button must be activated if you intend to transmit. Depending on how you've set up *WSJT-X*, it will become active automatically

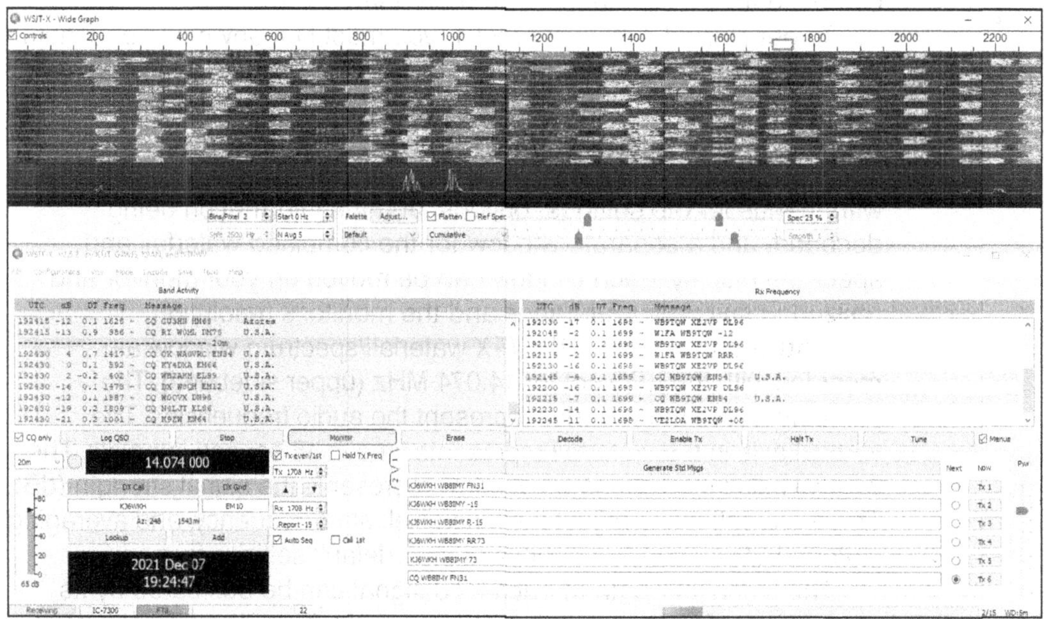

Figure 3.2 — Waterfall and spectrum views of FT8 activity on 20 meters.

when you call CQ, or answer someone else's CQ. Should you ever need to cease transmitting abruptly, just click your mouse cursor on the Halt Tx button.

Navigating the Settings Menus

Before you can use *WSJT-X*, you need to set up the software. You begin at the upper left FILE menu option and choose SETTINGS (or just press the F2 keyboard shortcut).

As you can see in **Figure 3.3**, you will need to enter MY CALL: and MY GRID: for your station. If you don't know your grid square, Dave

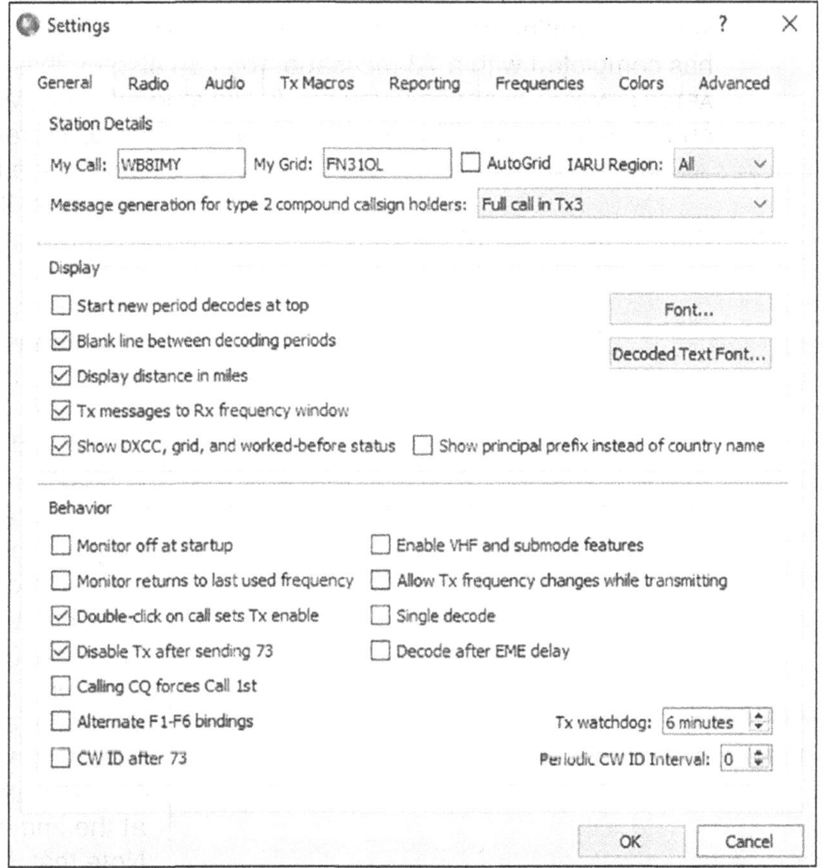

Figure 3.3 — Enter your call sign, grid square in the GENERAL window. You can set other program behaviors in this window as well. For example, be sure to check the box labeled Double click on call sets TX enable. **This will ensure that *WSJT-X* is in the transmit mode when you select a station calling CQ.**

Levine, K2DSL, has a handy calculator at **www.levinecentral.com/ham/grid_square.php**.

Under DISPLAY, check the first three options and under BEHAVIOR check the third and fourth options. An important feature in this window is the FONT and DECODED TEXT FONT buttons, which you can click on to match the size and type of the font you prefer.

Under BEHAVIOR, check DOUBLE CLICK ON CALL SETS TX ENABLE. That way, you do not have to constantly hit the ENABLE TX button whenever you want to transmit, such as when answering a CQ. The instructions in the rest of this chapter will be written under the assumption that you've checked the DOUBLE CLICK ON CALL SETS TX ENABLE box.

Also check DISABLE TX AFTER SENDING 73, which is a safety valve of sorts, preventing your continual transmissions even after your contact has completed with a 73 message. You can also enable the CW ID AFTER SENDING 73 option if you would like to identify in CW after the end of your contact. This is not needed, and you will rarely hear CW among the FT8 signals.

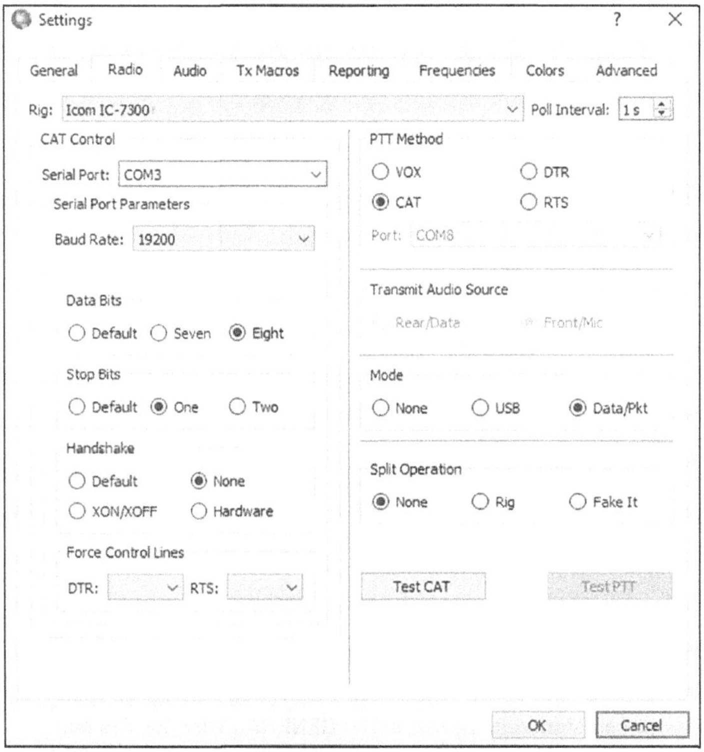

Under the RADIO tab, as shown in **Figure 3.4**, you can pick the method of switching your radio to the transmit mode. You'll find it in the PTT METHOD box. If you pick CAT control, you must choose the correct communication protocol for your transceiver (by using the drop-down "Rig" menu at the top), the serial (COM) port number, the COM port data rate, and the parity values for your rig in the CAT CONTROL box at the immediate left. Note that if you activate CAT without your radio turned on, or if *WSJT-X* is unable to communicate

Figure 3.4 — Set your CAT control (if any) and PTT transmit/receive control in the Radio window.

with your radio because of an incorrect COM port or speed setting, the program will generate an error message and you won't be able to continue.

Choose VOX (VOICE OPERATED SWITCH) in the PTT METHOD box if you are using an interface with VOX switching that responds to audio from your computer, or if you intend to allow your transceiver to switch from receive to transmit with its own VOX circuit.

Some transceivers will do transmit/receive switching (PTT) through the CAT connection. If your radio is one of these, select CAT as your PTT option. Otherwise, select either RTS or DTR and a COM port number for your keying device. In most cases this will be one of the interfaces we discussed earlier in this book. There is a TEST PTT button and TEST CAT button that you can use to be sure your PTT and/or CAT circuit is working.

CAT control is not necessary to operate *WSJT-X* in any mode, but it is convenient. For example, when you use the drop-down menu in the main window to select your operating frequency, your transceiver will switch to that frequency automatically. The software will also automatically pass your frequency to online reporting sites such as PSKReporter.

The next step, shown in **Figure 3.5**, is to select the AUDIO tab and choose the SOUND CARD device, both INPUT (receive) and OUTPUT (transmit). In this case, the transceiver is an Icom IC-7300 with its own built-in interface. The software

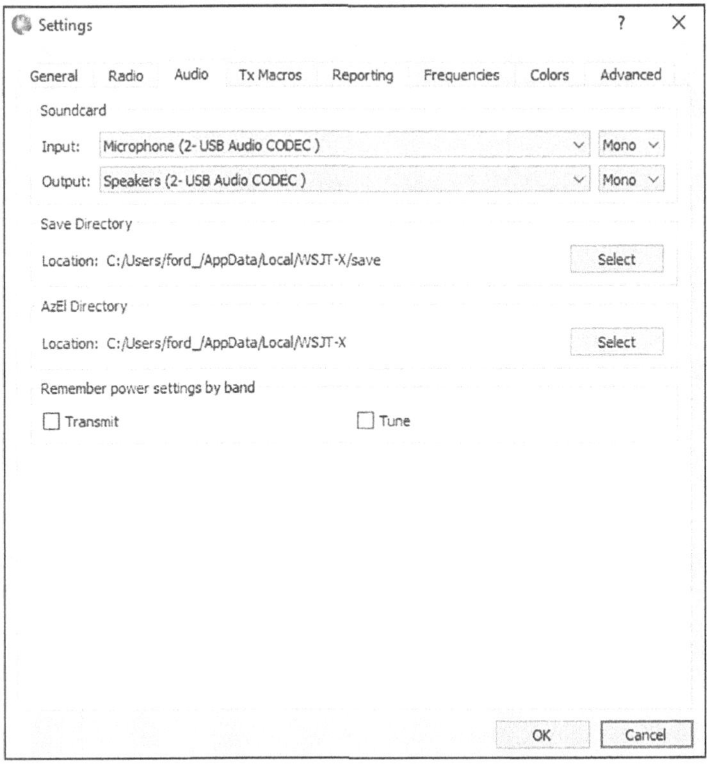

Figure 3.5 — Use the Audio tab to select your audio input and output options.

identifies the inputs and output as Microphone (USB Audio Device) and Speakers (USB Audio Device) respectively. If you are using an external interface with its own audio device, you will have to specify the COM port numbers your computer is using to "talk" to that device.

Figure 3.6 shows the REPORTING tab where you choose LOGGING QSO choices and to report the stations you copy to the PSKReporter web page (**https://www.pskreporter.info/pskmap**). This is a valuable tool that keeps track of digital mode signals received on all bands. If you have an internet connection at your station computer, be sure to check the ENABLE PSKREPORTER SPOTTING box.

When you exit the setup mentions and return to the main window, you must pick the MODE (FT8) from among the choices in *WSJT-X* and pick the DECODE speed. "Normal" usually works well. If you are blessed with a fast computer, you can select DEEP, which will allow you to decode weaker signals. However, this takes more time and could delay the speed at which the results appear on your monitor.

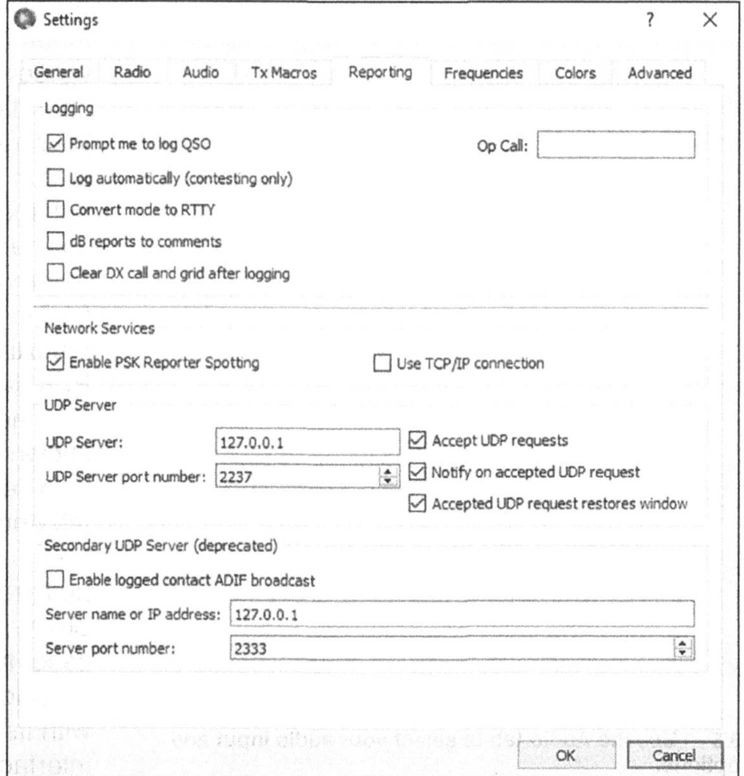

Figure 3.6 — Logging and PSKReporter options are set in this window.

The Importance of Time

In every *WSJT-X* operating mode, a critical component is time. The software depends on tight time synchronization to "know" when to expect data signals to be present and to respond accordingly. Your computer clock doesn't need to match the clocks of other stations exactly, but it must be synchronized within less than 2 seconds — the closer the better. If your computer clock is too far out of sync, you will not be able to decode signals.

When you look at decoded signals, pay attention to the numbers in the DT column. If you notice that most of the numbers are greater than 1, that's a clue that your clock computer is drifting out of sync with the other stations.

Fortunately, *Windows* has built-in time calibration that synchronizes automatically with time sources via the internet. Synchronization doesn't take place as often as it should and, as a result, *Windows* time may drift by a second or so after just a few days.

It is best to periodically "force" *Windows* to synchronize. In *Windows 10* or *11*, for example, open Control PANEL and click on TIME & LANGUAGE. In the DATE & TIME screen, you will see a "switch" labeled SET TIME AUTOMATICALLY. Slide the switch to the OFF position, wait a few seconds, and then slide the switch back to the ON position. This will force *Windows* to synchronize itself. You don't necessarily need to do this before every operating session, but it doesn't hurt.

A more convenient alternative is the free time-synchronizing application *Dimension 4*, which you can download at **www.thinkman.com/dimension4/**. Install the application and set it up so that it loads and runs constantly in the background whenever you start your computer.

Table 3.1
FT8 Frequencies
All frequencies are Upper Sideband (USB)

Band (Meters)	*Frequency (MHz)*
160	1.840
80	3.573
40	7.074
30	10.136
20	14.074
17	18.100
15	21.074
12	24.915
10	28.074
6	50.313

On the Air with FT8

With the menu items set and saved, the next thing to do is to pick a frequency and listen for activity. See the frequency list in **Table 3.1**. Switch your transceiver to upper sideband (USB, or USB-Data) and tune to that frequency. Don't touch the VFO knob again. If you have CAT enabled with your transceiver, *WSJT-X* will switch your radio to the correct frequency automatically when you select it from the drop-down menu located directly above the receive audio level display.

Do *not* choose a narrow passband 500 Hz filter if your radio offers one. Keep your IF passband set for a normal phone conversation, 2 to 3 kHz. This will allow you to see every signal that is scattered throughout the receive passband.

Having said that, there are times when you may want to use a narrower bandwidth. For example, if there is a station in the passband that is extremely strong, its signal may cause your AGC to reduce your receiver gain dramatically, causing weaker signals to disappear. Switching in a filter may help block the strong signal.

Never use speech processing, and never turn on noise blankers, or noise-reduction functions. They will distort your transmit signal and receive signals respectively. Also, if your transceiver has an AGC adjustment, set it to "fast."

Set your transceiver for maximum RF output (typically 100 W for many radios). Start by selecting an empty area of the waterfall display at about 1000 Hz and right-clicking the mouse. Now click on the Set TX RX Offset button that appears. You'll notice both the green and red brackets will move to your chosen spot. Next, click on the Tune button in the *WSJT-X* main window. If you've set up everything correctly, *WSJT-X* should key your transceiver and transmit a continuous tone.

Quickly click your mouse on the PWR slide control at the lower right corner and drag it down. Watch your transceiver's RF power output and ALC metering. Keep dragging the PWR control down until you see no ALC indication, or until the ALC meter indicates that it is within the minimal zone.

Continue dragging the PWR control until your RF output drops to where you feel it should be. You definitely don't need 100 W, but how much is too much? Many FT8 enthusiasts say you should never transmit more than 30 W; others suggest even less. The answer really depends on your antenna system. If you are using a gain antenna such as a Yagi, 30 W or less would be appropriate. On the other hand, if you are using a poor antenna such as an HF mobile whip sitting in your back yard with a couple of radial wires in the grass, you may need more than 50 W to be effective. Start with low power and see what sorts of signal reports you receive. If you consistently receive weak reports from most stations, that's a strong hint that you should increase power.

After tuning up, you are ready to operate, but before you make your first contact, start by just listening and observing. The first thing you'll notice is that transmissions take only 13 seconds, with a two-second pause between transmissions. This means the length of each transmit or receive cycle is a total of 15 seconds.

As you observe the exchanges, you'll see a pattern emerging. FT8 exchanges usually, though not always, follow a strict sequence. It goes like this, starting at 21:02:00 UTC. Note that when there is an exchange between two stations, the transmitting station's call sign is the one on the ***right***.

21:02:00 CQ WB8IMY FN31

WB8IMY sends CQ from grid square FN31.

21:02:15 WB8IMY N1NAS EN72

N1NAS replies and tells WB8IMY that he is in grid square EN72.

21:02:30 N1NAS WB8IMY -11

WB8IMY replies with a signal report of -11 dB.

21:02:45 WB8IMY N1NAS R-15

N1NAS acknowledges the signal report from WB8IMY with an "R" followed by a signal report (-15).

21:03:00 N1NAS WB8IMY RRR

WB8IMY sends "RRR," which means "Roger, roger, roger." Everything has been received and the exchange is complete.

21:03:15 WB8IMY N1NAS 73

N1NAS sends 73 – best wishes.

21:03:30 N1NAS WB8IMY 73

WB8IMY sends his 73 as well. The contact has ended.

The entire contact is complete within just 90 seconds, start to finish. Of course, this assumes that all goes well. If there is noise, fading, or interference that corrupts reception, an exchange may need to be repeated, and that will take longer. Under decent signal conditions, however, most contacts are completed in less than two minutes.

You may also see many stations skipping the RRR response and simply sending RR73 to indicate that the contact is complete.

Making Contact

If you're comfortable monitoring FT8, it is time to try a contact. Watch for someone calling CQ. Depending on how you've set up your colors in *WSJT-X*, a CQ may appear as green-shaded text. Make sure you've checked the Auto sequence (AUTO SEQ) box.

When you see a CQ, quickly double click your mouse on the CQ text. *WSJT-X* should respond by keying your transceiver and sending your call sign and grid square at the next opportunity. Now just sit back and watch; *WSJT-X* will handle the rest.

Automatic operating depends on *WSJT-X* receiving the correct text at the correct times. If it does, it will select the correct response and send it automatically. You won't need to lift a finger. When the station responds to you, *WSJT-X* should respond in turn.

Sometimes, however, things do not go as planned. The station may send a signal report but will not receive your confirmation and your report. Perhaps there was interference, or maybe conditions suddenly changed. If so, you may see the station repeating its report of your signal, and *WSJT-X* responding by sending your report of the other station's signal again. The other station's *WSJT-X* is doing this automatically because it didn't decode your text the first time around. Just wait as *WSJT-X* tries again. With luck, the station will receive and decode properly on the next attempt.

Once the contact is complete, usually with a 73 message, you'll notice that the Enable Tx button is no longer highlighted. You may also receive a prompt to add the contact to your *WSJT-X* log if you've enabled that in the settings menus.

Try Calling CQ

After you've made a few contacts, try calling CQ yourself. Once again, check the Auto Seq box, but this time you need to also check the CALL 1ST box to its immediate right. The CALL 1ST box is important because it lets *WSJT-X* "know" that you want to make a call (your CQ) at the next available opportunity without waiting for someone else's transmission to trigger it.

With each station copying all the signals in the passband, it does not matter where you call, so pick an open frequency where no one else is transmitting. You can use your mouse on the *WSJT-X* waterfall to select a spot to transmit by placing the cursor on the spot and right-clicking the mouse. Now click on the Set TX RX Offset button that appears. You'll notice both the green and red brackets will move to your chosen spot. Put a check the box labeled Hold Tx Freq so that your transmit frequency remains fixed while the receive frequency can switch if necessary to match someone who may be answering on a different frequency.

Now look in the GENERATE STD MSGS window and click your mouse in the circle on the bottom line, just to the left of the TX 6 button, which should be your CQ message. Click the ENABLE TX button and *WSJT-X* should begin transmitting at the next opportunity. If you happen to begin at the middle or near the end of a transmit cycle, the first transmission may be short. *WSJT-X* will likely wait until the *next* cycle is complete before sending the CQ transmission again.

During receive cycles *WSJT-X* will be "watching" for text that contains your call sign, the call sign of another station, and a grid square — in other words, a reply! When this occurs, you'll see the reply highlighted in red in the right-hand window. When that happens, *WSJT-X* will take control and attempt to complete the contact.

If no one replies, *WSJT-X* will send CQ again. When someone does reply, it will automatically go through the "dance" of completing the contact while you relax and watch.

Chasing Split-Frequency DX

Many DX stations use FT8 to rack up contacts. With that in mind, don't be surprised if you see DX stations calling at audio frequencies below 900 Hz on your waterfall display. These stations are not expecting you to answer on their frequency, though. Instead, they want to you transmit somewhere *above* 1000 Hz. As we discussed in the previous section, you'll need to pick a reasonably clear area above 1000 Hz, right-click your mouse on that spot, and then click on the SET TX RX OFFSET button that appears. Again, you'll notice both the green and red brackets will move to your chosen spot. Put a check the box labeled HOLD TX FREQ because once you have chosen your transmit frequency, you don't want it to change to the frequency of the DX station. Checking this box will ensure your transmitting frequency won't change automatically when you double-click on the DX station's CQ text.

When you double click on the DX station's CQ, *WSJT-X* will automatically change your receive frequency to match their transmitting frequency, but your transmitting frequency will remain unchanged. This may seem counterintuitive, but don't worry. No matter where you are transmitting, if that frequency is clear and your signal is sufficiently strong, the DX station will see your response.

Fox and Hound Mode

When DXpeditions are operating FT8 or FT4, things can get very intense. That's why they will often use the *Fox and Hound* mode.

Fox and Hound is a special *WSJT-X* feature designed for use by DXpeditions where one station is attempting to make contacts with stations as rapidly as possible. DXpedition operators are discouraged from using Fox and Hound mode at the usual FT8/FT4 frequencies because the crush of signals can be very disruptive to everyone else. Instead, they will choose frequencies well beyond the usual FT8/FT4 watering holes. On 20 meters, this might be 14.095 MHz, for example. They will typically announce their frequencies on DX Clusters or on DXpedition websites.

The DX station is the Fox, and it transmits at audio frequencies between 300 and 900 Hz. When transmitting multiple simultaneous signals, the signals are spaced at 60 Hz intervals. Hounds (that's you) make initial calls anywhere in the range 1000 – 3000 Hz. The Fox will not respond to Hounds calling below 1000 Hz.

To join the fray as a Hound, you'll first need to go into the *WSJT-X* settings menu and click on the Advanced tab. Put a check in the box labeled Special operating activity, and then click on the Hound circle.

When you return to the main *WSJT-X* screen, you should see a small red indicator that you are in a SPECIAL OPERATING MODE. Next, you need to choose the operating frequency. Since it will typically be a non-standard frequency, you will need to enter it by typing directly in the "Band/Frequency" box. Highlight the band information in the "Band/Frequency" box, type in the desired frequency and then press ENTER to accept the typed frequency. The background color frequency box will turn red to indicate a non-standard frequency. Using the instructions we discussed earlier, select a position on the waterfall above 1000 Hz and make that your transmitting spot. Try to pick a spot that is clear of other signals.

Now listen/watch for the Fox station to be decoded. Once you see the Fox being decoded, double click on its CQ message to begin transmitting. The transmit option will only include "Odd" intervals, so the Even/first option will be grayed out.

Your station (and many others) will begin calling and the *WSJT-X* software running at the DXpedition location will create a list of several calling stations and the operator will answer them in sequence. You don't need to do anything except wait and hope you are one of the stations in the list. You'll know you've made the list when you see the DX station calling you and your *WSJT-X* software sending its reply. The Fox will send your signal report, await your response, and then end with **RR73**. It will send your signal report no more than three times without hearing your response before it gives up and goes to the next station in the list. If the DX starts calling CQ again, it means you didn't make it this time around. Double click on the station's CQ and try again.

Always remember to uncheck SPECIAL OPERATING ACTIVITY to return to normal operations. For a much more detailed explanation, read the FT8 DXpedition Mode document at **physics.princeton.edu/pulsar/k1jt/FT8_DXpedition_Mode.pdf**.

A Few Words about FT4

WSJT-X added the FT4 mode to provide faster operating during contests. Some DXpeditions will use FT4 in Fox and Hound mode to maximize their contact rates. When it comes to operating, the procedures are the same as FT8, but you can also make use of special contest modes in the ADVANCED tab of the *WSJT-X* settings menu.

The FT4 signal protocol is different, and it sacrifices a bit of gain for the benefit of enhanced speed. The margin isn't large, but it is

Table 3.2
FT4 Frequencies
All frequencies are Upper Sideband (USB)

Band (Meters)	*Frequency (MHz)*
40	7.0475
30	10.140
20	14.080
17	18.104
15	21.140
12	24.919
10	28.180
6	50.318

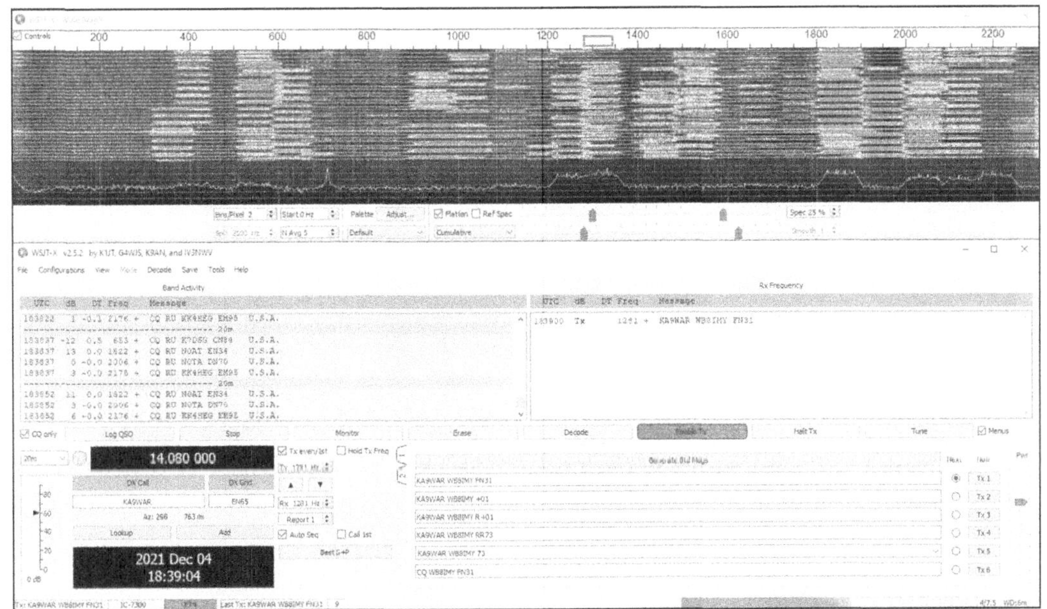

Figure 3.7 — A busy FT4 display during a weekend contest. Notice how different the waterfall appears compared to FT8. This is due to the fact that the individual transmission/reception cycles are much shorter in duration.

enough to potentially make a difference if you are attempting to contact a weak station.

Table 3.2 shows the most common FT4 frequencies. When this edition was written, FT8 was the most popular of the two modes by a wide margin and the contest community had yet to adopt FT4 in large numbers. Despite this, you'll still find hams making casual FT4 contacts every day. See **Figure 3.7**.

Chapter 4

PACTOR

Hans-Peter Helfert, DL6MAA, and Ulrich Strate, DF4KV, developed PACTOR in 1991. There are now four versions of PACTOR: PACTOR I, PACTOR II, PACTOR III, and PACTOR IV. PACTOR I is little used. PACTOR II and III are in current use, primarily to connect to automated stations such as the Winlink network. See the Appendix for a description of Winlink.

There is no sound-device software method for sending and receiving PACTOR. As mentioned in Chapter 1, the only way to get on the air with PACTOR is to use a stand-alone hardware *controller* (also referred to as a *modem*) that connects between your computer and an SSB transceiver.

We call PACTOR a burst mode because of the way it sends information. Rather than sending a continuous stream of data like RTTY, PACTOR transmits bursts of information that take the form of short data blocks. When the data is received intact, the receiving station sends an ACK signal (for acknowledgment). If the data contains errors, a NAK is sent (for nonacknowledgment). In simple terms, ACK means, "I've received the last group of characters okay. Send the next group." NAK means, "There are errors in the last group of characters, send them again." This back-and-forth data conversation sounds like crickets chirping. In the case of PACTOR, the long chirp is the data, and the short chirp is the ACK or NAK. A digital mode that functions in this manner is said to be using *ARQ* – Automatic Repeat Request.

Memory ARQ

In less sophisticated burst modes such packet, a data block must be repeated over and over if that's what it takes to deliver the information error-free. This results in slow communication, especially when conditions are poor.

PACTOR manages the challenge of "repairing" errors in an interesting way. Each data block is sent and acknowledged with an ACK signal if it's received intact. If signal fading or interference destroys some of the data, a NAK is sent, and the block is repeated. Nothing new so far — packet radio behaves in much the same way. The substantial difference, however, involves memory.

When a PACTOR controller receives a mangled character block, it analyzes the parts and temporarily memorizes whatever information appears to be error-free. If the block is shot full of holes on the next transmission as well, the controller quickly compares the new data fragments with what it has memorized. It fills the gaps as much as possible and then, if necessary, asks for another repeat. Eventually, the controller gathers enough fragments to construct the entire block. PACTOR's memory ARQ feature dramatically reduces the need to make repeat transmissions of damaged data. This translates into much higher throughput.

PACTOR has the capability to communicate at varying speeds according to band conditions. PACTOR throughput is enhanced by using *Huffman coding* that reduces the average character length for improved efficiency.

What Do You Need to Operate PACTOR?

Assembling a PACTOR station is quite simple. All you need are the following:

• **An SSB transceiver.** The output of the PACTOR controller is audio, which can be applied at the microphone or accessory jack. The controller also operates the transmit/receive keying line. The only catch is that the transceiver must be capable of switching between transmit and receive in less than about 20 milliseconds. Most modern transceivers meet this requirement. If you're unsure, check the *QST* magazine Product Reviews. The ARRL Lab routinely tests this function in every transceiver review.

• **A controller with PACTOR capability.** First-generation controllers by Timewave and older Kantronics and AEA controllers

The SCS DR-7400 P4Dragon controller supports all PACTOR modes (including PACTOR IV) as well as packet, RTTY, and more.

support PACTOR I, but only controllers manufactured by Special Communications Systems (SCS) offer PACTOR II, III, and IV.

• **A computer running the controller's companion software.** Most compatible software is written for Microsoft *Windows*. For amateur radio use, the most popular application is *Winlink Express*, which allows you to use the PACTOR controller to access the Winlink global email network. We'll discuss *Winlink Express* in more detail in Chapter 6.

When assembling your PACTOR station, use the guidelines discussed in Chapter 1. The connection to the computer is made with a serial cable in older controllers, so you may need a USB/serial adapter if your computer doesn't have a serial port. The latest SCS PACTOR units offer a USB connection.

CQ PACTOR

Although most PACTOR activity involves connecting to automated systems, you can still occasionally enjoy a "live" conversation if you can connect to the controller using terminal software such as the free *Alpha* application available at the SCS website at **www.p4dragon.com/downloads.html**. The controller manual lists all the available user commands.

You call CQ on PACTOR using *forward error correction*, or FEC. The FEC signal sounds like fast RTTY, but in reality, it is a stream of data in which each character is repeated twice for redundancy--there are no ACKs or NAKs. Obviously, this means that an FEC transmission is not error-free, but the copy is good enough to pull out the call sign of the sending station. When you're sending a PACTOR CQ, that's really all that matters.

If you hear a signal that you suspect is a PACTOR FEC CQ, tune carefully until your controller indicates that it has locked (synchronized) with the FEC signal. Within a short time, you should begin to see text on your screen.

CQ CQ CQ DE WB8IMY WB8IMY WB8IMY
CQ CQ CQ DE WB8IMY WB8IMY WB8IMY
CQ CQ CQ DE WB8IMY WB8IMY WB8IMY K K

Notice that this CQ uses several short lines of text rather than a few long lines. This helps stations synchronize more easily.

Starting a Conversation

Answering a CQ in PACTOR is straightforward. Depending on the software you're using, it may be as simple as entering:

CALL N1BKE, or
CONNECT N1BKE

at the CMD: prompt. Some types of software streamline the process even further. There may be pop-up boxes where you simply enter a call sign.

Once you make contact, the conversation proceeds in turns. This means that one station "talks" (the ISS or information sending station) while the other "listens" (the IRS or information receiving station). When the ISS has finished, the sending station sends a special control signal known as the *over command* that immediately reverses the roles — suddenly you are the ISS and the other person is the IRS. Introduce yourself and ask a question about where the receiver lives or what the receiver does for a living. Use the over command to flip the link again. A conversation is underway!

Depending on the software you are using with your controller, sending the over command is usually as easy as tapping a single key. The software usually provides some sort of visual indicator to show which mode you are in — IRS or ISS — in case you become confused! If you forget to send the over command, your stations will simply sit there and chirp mindlessly at each other. Fortunately, the IRS can send an over command and flip the link for you if necessary. This is known as a "forced over."

PACTOR II

You can think of PACTOR II in terms of being a supercharged version of PACTOR I. In good conditions, PACTOR II boasts a data transfer rate up to six times faster than PACTOR. At the same time, PACTOR II can maintain links in conditions where the signal-to-noise ratio is at –18 dB. This means that PACTOR II can continue an exchange even when the signals are virtually inaudible. Such remarkable performance is also conservative when it comes to bandwidth; a PACTOR II signal only occupies about 500 Hz of spectrum.

The complex pi/4-DQPSK modulation system used by PACTOR II requires DSP technology and fast microprocessors. With DSP, you let the software decide the composition of the signal, not the hardware. DSP is much more flexible, allowing you to create the signal you want without having to worry about hardware filters and so on. The tradeoff is that you must use high-speed microprocessors to handle all the incoming and outgoing information at a decent rate. That's part of the reason PACTOR is implemented by SCS in dedicated hardware controllers.

PACTOR II is also backward compatible with PACTOR I. That is, a PACTOR II operator can communicate with a PACTOR I operator and vice versa.

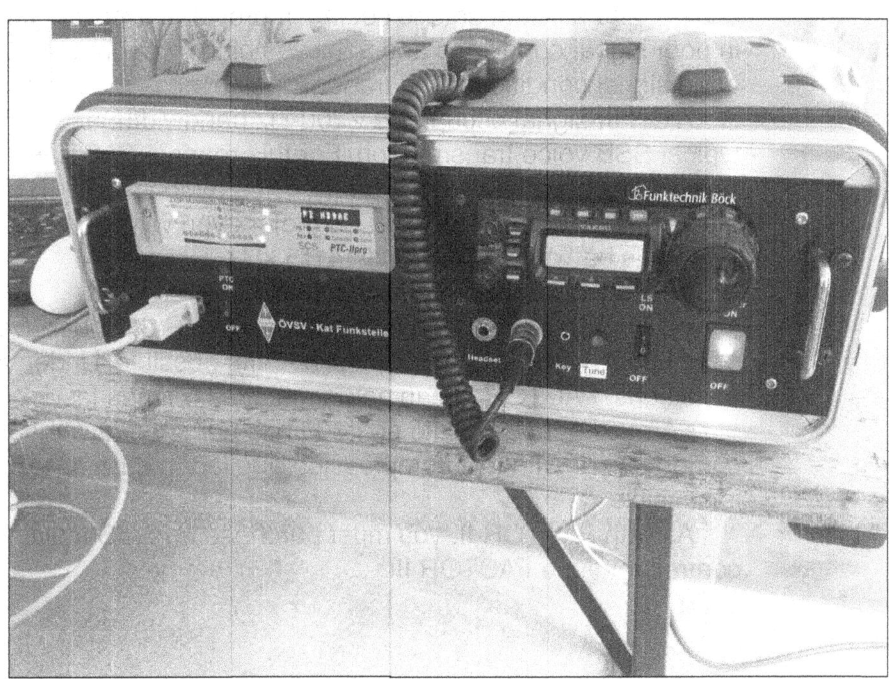

A PACTOR controller housed in an emergency communication go-kit along with an HF transceiver.

PACTOR II Station Requirements and Operation

The requirements for a PACTOR II station are essentially the same as for a PACTOR I station: An SSB transceiver, a computer, and an SCS modem.

PACTOR II also operates in nearly the same fashion as PACTOR I. In fact, the link is initially established using the PACTOR I protocol and then automatically switched to PACTOR II if both stations are using PACTOR II processors. And as with PACTOR I, you must take turns during the conversation, sending an over command to allow the other operator to send data.

PACTOR III

PACTOR III was introduced by SCS in 2001 as a firmware upgrade to their multimode controllers. PACTOR III is used primarily in the Winlink network, especially among network stations in the United States.

PACTOR III represents a substantial improvement with its 2-dimensional orthogonal pulse shaping, advanced error control coding, and efficient source coding. It provides excellent performance in poor signal conditions and achieves high throughput rates under good signal conditions. The tradeoff, however, is the fact that a PACTOR III signal occupies a 2200-Hz bandwidth, as much as a typical SSB voice transmission. For this reason, PACTOR III stations often operate well above the frequencies used for FT8, RTTY, and other modes. For example, on 20 meters you will often hear PACTOR III signals between 14103 and 14115 kHz.

The key benefits of PACTOR III are:

• Higher throughput compared to PACTOR II. Under average signal conditions, a speed gain factor of 3 – 4 is possible. Under very good conditions, PACTOR III can be as much as 5 times faster.

• A maximum data rate of 2,700 bits per second without compression. With text compression applied, a data rate of 5,200 bps is possible.

As with PACTOR II, you must purchase an SCS multimode controller to use PACTOR III.

PACTOR IV

SCS unveiled PACTOR IV in early 2011. The significant difference between PACTOR IV and PACTOR III is speed. The PACTOR IV protocol boasts a symbol rate of 1800 baud within a 2400 Hz bandwidth using 10 speed levels, DBPSK/DQPSK (non-coherent, spreading factor 16), BPSK-32QAM (coherent) and adaptive equalizing. All this translates into astonishing throughput that is potentially more than double that of PACTOR III. Squeezing that kind of HF digital performance into a 2400-Hz bandwidth required years of painstaking work.

Unfortunately, PACTOR IV is not legal for amateur use in the United States. Below 28 MHz American amateurs are restricted to data modes with effective symbol rates of 300 baud or less and PACTOR IV exceeds this limit substantially.

When natural disasters occur, ARRL often makes requests to the Federal Communications Commission to temporarily allow the use of PACTOR IV. These requests are almost always granted because the FCC is aware of PACTOR IV's ability to transfer critical data in and out of disaster areas at substantial speeds.

At the time of this writing, the ARRL has requested that the FCC consider a permanent modification of the Part 97 rules that would effectively allow higher baud rates on the HF bands. Should this come to pass, PACTOR IV would become legal for US amateurs on a regular basis.

Chapter 5

WSPR

WSPR is an acronym for Weak Signal Propagation Reporter. Most hams simply refer to it as "whisper." Unlike the other digital modes discussed in this book, WSPR does not support two-way communication. Instead, WSPR operates in a one-way transmission mode only.

WSPR signals can be received under conditions that would render any other digital signals unreadable. You can decode a WSPR signal that you literally cannot hear. An extremely weak signal may appear as little more than a faint, ghostly trace on a waterfall display, if it is visible at all.

WSPR was created by Dr. Joe Taylor, K1JT, in 2008, and today it is included in his *WSJT-X* software suite, which is available for *Windows*, *Linux*, and *MacOS* at **https://physics.princeton.edu/pulsar/k1jt/wsjtx.html**.

Each WSPR transmission carries the following information:

- Your call sign.
- Your Maidenhead grid square.
- Your RF output power (in dBm).

Each transmission is a little more than 110 seconds long and begins one second after the beginning of an *even* minute, such as 1202, 1204 and so on. If you hear a WSPR signal on the air, it will sound like a single tone that warbles ever so slightly. What sounds like an absence of modulation is a product of the fact that the WSPR symbol rate is less than 2 baud. A WSPR signal is also quite narrow at only 6 Hz.

WSPR is typically transmitted with less than 5 W RF output — sometimes much less. In fact, it is common to see WSPR transmissions taking place at milliwatt power levels. This may seem like an extraordinarily low power level, but WSPR doesn't require much energy to be heard over great distances.

Gathering Reports: WSPRnet.org

At the heart of the WSPR universe is a website known as **WSPRnet.org**. This website is a hub that gathers WSPR signal reports from stations throughout the world. At WSPRnet, you can see who is hearing your signals, and how well they are hearing them. You can display this information in the form of a colorful map or see it as columns of text data.

WSPRnet data is updated every couple of minutes, around the clock. You can sort the information and see only the stations that are hearing you on a given band. You can even download an enormous amount of WSPRnet data and process it using your favorite software.

It is fair to think of WSPRnet as a *crowdsourced* global information repository. Anyone who is operating WSPR with *WSJT-X* software can contribute to WSPRnet. All you have to do is select the WSPR mode in *WSJT-X* and then check the box labeled UPLOAD SPOTS. Even if you have only a low-speed internet connection at your station, that's more

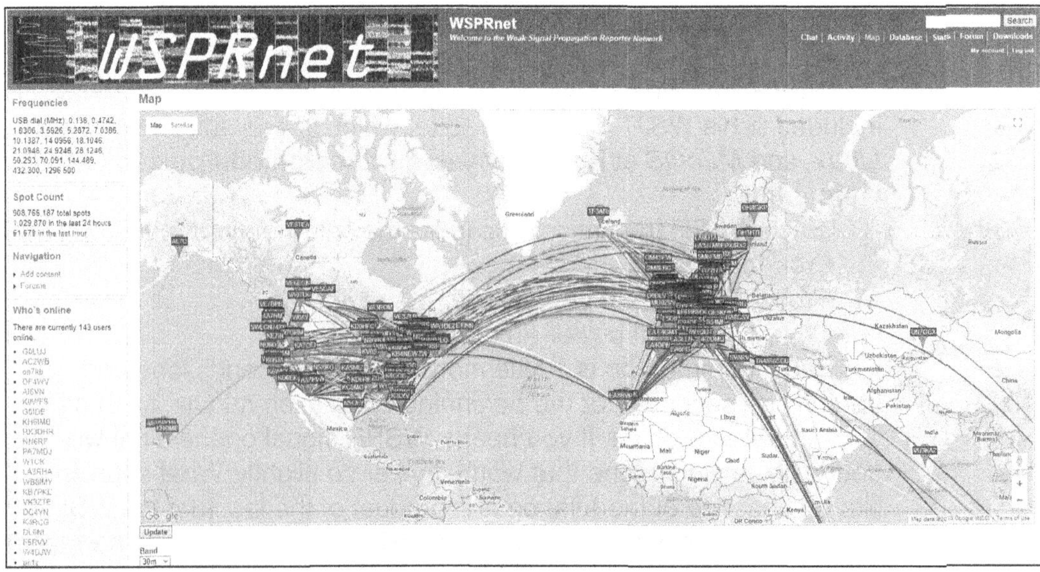

An amazing map display of WSPR reception reports on WSPRnet.org.

WSPRnet

Welcome to the Weak Signal Propagation Reporter Network

Chat | Activity | Map | Database | Stats | Forum | Downloads

My account | Log out

Frequencies

USB dial (MHz): 0.136, 0.4742, 1.8366, 3.5926, 5.2872, 7.0386, 10.1387, 14.0956, 18.1046, 21.0946, 24.9246, 28.1246, 50.293, 70.091, 144.489, 432.300, 1296.500

Spot Count

908,766,187 total spots
1,029,670 in the last 24 hours
51,678 in the last hour

Navigation

- Add content
- Forums

Who's online

There are currently 141 users online.

Database

Specify query parameters

50 spots:

Timestamp	Call	MHz	SNR	Drift	Grid	Pwr	Reporter	RGrid	km	az
2018-01-27 17:05	VK2TGQ	14.097178	-19	0	QF56of	5	VK2TGQ/1	QF56of	0	0
2018-01-27 17:04	UA1WCF	7.040140	-10	-2	KO55ir	10	PA0EHG	JO22hb	1746	267
2018-01-27 17:04	G4EWS	1.838084	-19	0	IO83	0.2	GW2HFR	IO83ib	46	180
2018-01-27 17:04	M1LCR	7.040158	-8	0	IO82rb	5	EA1IOW	IN83gj	966	184
2018-01-27 17:04	OZ7IT	3.594156	-4	0	JO65df	6	SA6BSS/RX	JO68sc	329	13
2018-01-27 17:04	WA4KFZ	7.040135	-10	0	FM18gv	5	NV0O	EM28tt	1463	275
2018-01-27 17:04	DL0PBS	7.040000	-16	0	JO33	0.2	OH8GKP	KP24rp	1619	33
2018-01-27 17:04	EA7GO	7.040126	+1	0	IM87ds	0.01	PA0EHG	JO22hb	1717	20
2018-01-27 17:04	VE3RQX	14.097125	-23	-1	FN03kt	0.01	K4MSC	EM90fo	1484	189
2018-01-27 17:04	EA7GO	7.040092	-1	0	IM87ds	0.01	PA0KNW/1	JO33md	1899	22
2018-01-27 17:04	W8AC	7.040031	-10	-1	EN91jm	0.001	VE3YX	FN16fd	592	29
2018-01-27 17:04	N8YE	7.040087	-26	0	EM99hj	1	KC8R	EN82lq	392	340
2018-01-27 17:04	ON5VW	5.366249	-18	0	JO10vv	1	OH8GKP	KP24rp	1977	31
2018-01-27 17:04	UA1WCF	7.040134	-8	-3	KO55ir	10	DG4DW	JO31so	1585	263
2018-01-27 17:04	G1ZPJ	7.040106	-13	0	IO70ic	0.2	PA0EHG	JO22hb	725	69
2018-01-27 17:04	AC6OT	14.097022	-16	4	CM87	10	KI6STW	EL98lo	3980	92
2018-01-27 17:04	AC0G	14.097075	-8	-3	EM38ww	1	KI6STW	EL98lo	1522	136
2018-01-27 17:04	EA7GO	7.040119	+8	0	IM87ds	0.01	S54MI	JN65un	1681	53
2018-01-27 17:04	WA4KFZ	7.040135	-6	0	FM18gv	5	KN8DMK	EM89oo	466	281
2018-01-27 17:04	DL7JKD	7.040152	-22	0	JO62ek	2	G0NJS	IO91vs	941	271
2018-01-27 17:04	EA5CYA	7.040065	-24	0	IM99	0.2	ON7KB	JO21el	1386	16
2018-01-27 17:04	VE3SAO	14.097118	-23	0	EN58jk	0.2	K1HEK	CN86tt	2521	279
2018-01-27 17:04	WH2XXP	0.137507	-18	0	DM33	1	W7IUV	DN07dg	1633	342
2018-01-27 17:04	LA9DTA	5.288709	-25	0	JP50	0.2	OH8GKP	KP24rp	873	52
2018-01-27 17:04	9A5PH	7.040103	-11	0	JN73tt	1	PA0EHG	JO22hb	1227	322
2018-01-27 17:04	VA3UAL	7.040149	-18	0	EN94xc	0.6	NV0O	EM28tt	1329	249
2018-01-27 17:04	KK6I	14.097108	-16	0	DM13tb	5	VA7IS	CN89mh	1872	347
2018-01-27 17:04	N8GQM	14.097133	-22	0	EN91	0.1	K4MSC	EM90fo	1210	182
2018-01-27 17:04	OZ2JBR	5.268689	-9	0	JO65dl	5	OH8GKP	KP24rp	1261	30
2018-01-27 17:04	WA4KFZ	7.040138	-8	0	FM18gv	5	VE3RPH	FN03hq	556	344
2018-01-27 17:04	SP3JIA	7.040006	-11	0	JO82kk	1	PA0EHG	JO22hb	834	272
2018-01-27 17:04	G0ORD	7.040059	-17	0	IO81ru	5	OE6WSF	JN76ts	1428	106
2018-01-27 17:04	G8WVW	7.040168	-24	0	IO81vu	0.2	EA1IOW	IN83gj	945	186
2018-01-27 17:04	WA4KFZ	7.040136	+14	0	FM18gv	5	KC8R	EN82lq	631	314
2018-01-27 17:04	N8YE	7.040088	+4	0	EM99hj	1	VE3YX	FN16fd	813	21
2018-01-27 17:04	KD2LZI	14.097092	-21	-1	FN22vt	0.2	KI6STW	EL98lo	1703	204
2018-01-27 17:04	OZ7IT	3.594152	-17	0	JO65df	5	OH8GKP	KP24rp	1273	29
2018-01-27 17:04	EI3AL	7.040076	-6	0	IO52ab	0.5	PA0EHG	JO22hb	961	64

WSPRnet.org will also display reports in text format.

than sufficient to upload information to WSPRnet. Uploading takes place behind the scenes and it won't interfere with the operation of the rest of the software.

Many *WSJT-X* users often set it up to receive in WSPR mode and then walk away for hours, or even leave the software running in receive-only mode overnight. It is a good feeling to know you're making a substantial contribution to the WSPR community when you are not otherwise using your station.

You don't need to register for an account at WSPRnet, but it is a good idea to do so. Registration is free and it gives you access to other WSPRnet features such as the website forum where you can ask questions and get assistance from experienced operators.

Three Ways to WSPR

There are three ways to enjoy WSPR:

• As a WSPR "transceiver," which means that you are sending as well as receiving. This requires an SSB transceiver, a computer, and *WSJT-X* software.

• As a receive-only station. As described previously, this means you are only receiving WSPR signals and presumably uploading reports to WSPRnet.

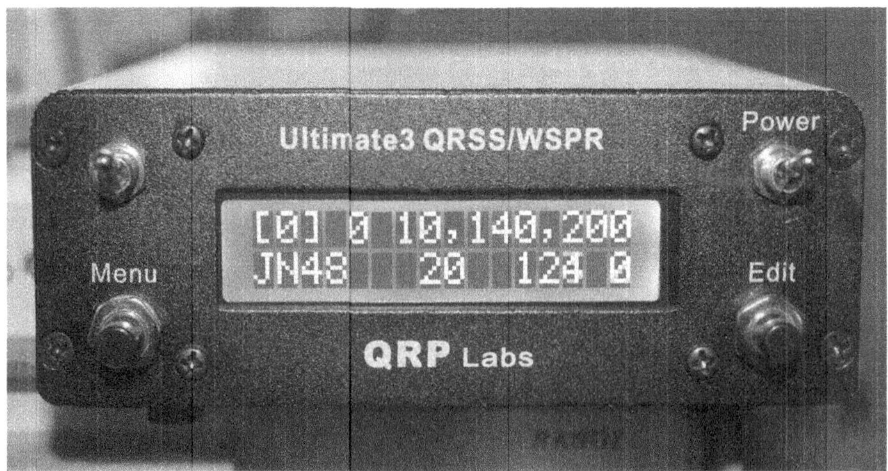

The QRP Labs Ultimate3 is a transmit-only beacon radio for WSPR and other modes. [K5ACL photo]

Tucson Amateur Packet Radio (TAPR) offers a plug-in board that will add a 20-meter WSPR transmitter to a Raspberry Pi microcomputer.

- As a transmit-only station. These stations are using transmitters dedicated to WSPR operation. For example, the QRP Labs Ultimate3 (**www.qrp-labs.com**) is a transmitter with firmware that allows it to function as a low-power beacon for WSPR and other modes. There are also the tiny WSPRLite transmitters by SOTABEAMS (**www.sotabeams.co.uk**) that function as low-power beacons for WSPR only. It is also possible to add modules to Raspberry Pi microcomputers and turn them into WSPR transmitters. Tucson Amateur Packet Radio (TAPR) sells an inexpensive module that creates a WSPR signal on 20 meters (**tapr.org/product/wspr/**) when paired with a Raspberry Pi.

WSPR with *WSJT-X*

In Chapter 3 we discussed how to set up *WSJT-X* software for FT8 and FT4. The same basic settings apply to WSPR. To use *WSJT-X* for WSPR specifically, you just click on the MODE menu and select WSPR.

Table 5.1 **WSPR Frequencies**	
Band (meters)	*Display Frequency, Upper Sideband*
160	1.8366
80	3.5926
40	7.0386
30	10.1387
20	14.0956
17	18.1046
15	21.0946
12	24.9246
10	28.1246
6	50.2930

When you select the WSPR mode, the *WSJT-X* window will change instantly to its WSPR configuration. If you have set up CAT control for your transceiver, your rig will probably switch to whatever WSPR frequency has been selected in the drop-down frequency/band menu. If you aren't using CAT, set your transceiver to one of the WSPR frequencies shown in **Table 5.1**. Also switch your transceiver to upper sideband (USB).

In the waterfall display, look for the thin, green bar along the top edge. This marks the receive audio bandwidth, which is a 200-Hz slice between 1400 and 1600 Hz. Above the green bar, you'll find a small, red marker. This designates your transmit audio frequency. Think of the red marker as your transmit point.

In the bottom portion of the main window, there are several items that need your attention. See **Figure 5.1**. In the lower left, you'll find the drop-down menu for your CAT control, as we discussed earlier. In Figure 5.1 you'll notice that it happens to be set for 30 meters. Below the band selection menu, you'll find the receive audio level control.

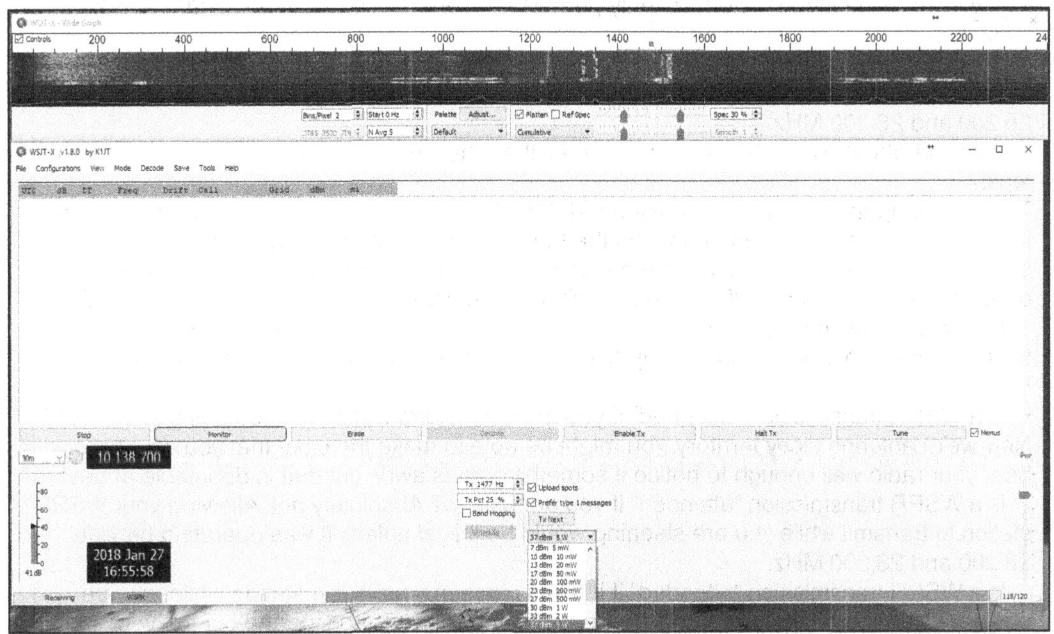

Figure 5.1 — *WSJT-X* in WSPR mode.

Set this for a happy medium between too little and too much audio, as shown in Figure 5.1.

In the middle, you will see two small spinner boxes. The top box indicates your transmit audio frequency. This corresponds to the position of the red marker above the waterfall display.

The bottom box is used to choose how often you want to transmit. Transmissions occur on a semi-random basis so that everyone has an opportunity to transmit without too much interference. If you select 25%, for example, you will only transmit on 25% of the available even minutes. At the opposite extreme, if you select 100%, *WSJT-X* will transmit on *every* available even minute. A setting of 25% is a good choice — enough transmissions to assure that your signal will be heard, yet not so many that your transmissions become obnoxious. If you need to force *WSJT-X* to transmit on the next available even

FCC Rules and WSPR

Part 97 of the FCC Rules permits amateurs to make one-way transmissions for testing purposes. It is this aspect of the rules that makes WSPR legal on the HF bands. "Testing," as it applies to WSPR, is testing of equipment (antenna systems) and propagation.

Beacons, by definition, are unattended stations that make one-way transmissions. The key word is "unattended." The FCC only permits beacons on HF between 28.200 and 28.300 MHz.

So, is a WSPR station a beacon station as defined in Part 97? If the transmitter is unattended, yes. That means that unattended WSPR beacons can only operate legally between 28.200 and 28.300 MHz.

But what about WSPR transmissions on all the other HF bands? Are these transmissions legal?

Yes, so long as the stations are attended. The FCC generally defines "attended" as meaning that someone is keeping an eye on the transmissions, either locally or remotely. The station doesn't need to be monitored continuously, but it must be monitored with enough frequency to ensure that a malfunction doesn't go undetected for an excessive amount of time.

Is a WSPR transmission "attended" if you are busy building a kit on the workbench next to your radio? Most people would say "yes." If something goes wrong, you're likely to notice quickly.

Is a WSPR transmission "attended" if you are downstairs watching a football game on TV? Now we're entering dicey territory. You might be able to make the case that you could still hear your radio well enough to notice if something goes awry, but that is debatable at best.

Is a WSPR transmission "attended" if you are asleep? Absolutely not. Allowing your WSPR station to transmit while you are sleeping would be illegal unless it was operating between 28.200 and 28.300 MHz.

Is a WSPR transmission "attended" if it takes place at your home station while you are away at school, or at the office? Again, no — unless you can monitor and control your station remotely.

Transmit-Only WSPR Rigs and Spectral Purity

As mentioned in this chapter, it is possible to build a dedicated WSPR transmitter with WSPR firmware controlling the RF circuitry, or with a microcomputer such as a Raspberry Pi that is running *WSJT-X* and using a module to generate RF output. These are usually very low-power rigs, but they can do the job.

One thing to beware of, however, is spectral purity. Some of these transmitters are little more than signal generators with antenna jacks. Their RF output tends to be remarkably "dirty" with many signals spraying across the radio spectrum.

To be legal under the requirements of the FCC Rules, the mean power levels of any spurious emissions must be at least 43 dB below the mean power level of the fundamental signal. Most of these dedicated WSPR rigs won't come anywhere close to being legal without well-designed low-pass filters in the output circuitry. When considering these transmitters, take a careful look at their specification and consider adding a low-pass filter if you are unsure.

minute, regardless of the transmit percentage you've selected, click on the nearby TX NEXT button.

To the right of the small spinner boxes, you'll find two checkboxes. UPLOAD SPOTS we've already discussed. Check this box so that your computer will upload your reception reports to WSPRnet. PREFER TYPE 1 MESSAGES is a box best left checked. The brief explanation is that a WSPR transmission can support 6-character grid square designations. However, these are almost never used on the HF bands. When you check the PREFER TYPE 1 MESSAGES box, you are telling the software not to bother attempting to send 6-character designators.

Below these boxes, you will see the drop-down box that allows you to specify your transmitted RF output. It is important to realize that this menu only selects the information to be included in your transmitted data — *it does not automatically set your RF output*. Regardless of what you select in this menu, you must still manually set your RF output by using the TUNE button and the vertical PWR slide control on the right.

When it comes to setting RF output for WSPR, most amateurs have RF power meters — either within their transceivers or as external units — that can read down to 5 W. Setting power levels below 5 W can be difficult because it requires a wattmeter that can measure and accurately display at lower power levels. Such a device isn't common in amateur stations. That's why many *WSJT-X* users default to 5 W (37

Table 5.2
Power and dBm

Power	*dBm*
5 mW	7
10 mW	10
20 mW	13
50 mW	17
100 mW	20
200 mW	23
500 mW	27
1 W	30
2 W	33
5 W	37

dBm) when operating WSPR. A level of 37 dBm translates to 37 dB above 1 milliwatt, or 5 W. See **Table 5.2**.

Below the center windows, you will see a checkbox labeled BAND HOPPING. This is a clever function included in *WSJT-X* that's ideal for studying propagation. If you have set up *WSJT-X* for CAT control of your transceiver, you can check this box and set up a schedule for WSPR transmitting and monitoring on the bands you desire. See **Figure 5.2**.

For example, you can check the boxes to make *WSJT-X* automatically jump from 20 meters in the daytime to 80 meters at night. Or it can hop from 20, to 15, to 10 meters during the daylight hours only.

Receiving

Once you have *WSJT-X* configured to your liking for WSPR, make sure the MONITOR button is green (click on it) so that *WSJT-X* will begin monitoring. Either tune your radio to a WSPR frequency or use the drop-down menu and let CAT do it for you. Now just sit back and wait, or busy yourself with another task. Give *WSJT-X* some time to listen for WSPR signals.

If there are WSPR signals present, you will soon see that *WSJT-X* has decoded several of them. These will appear in the middle section of the display. See **Figure 5.3**.

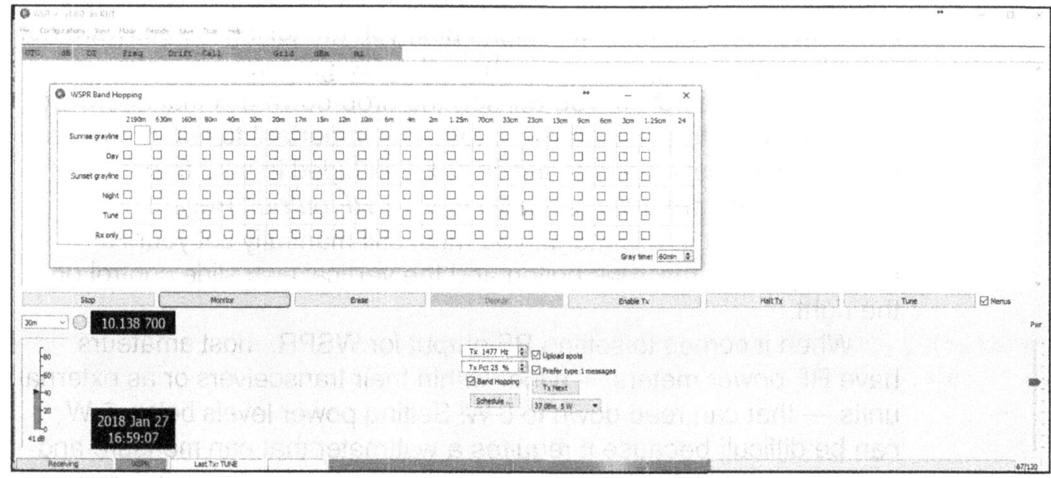

Figure 5.2 — You can configure *WSJT-X* in WSPR mode to automatically hop from one band to another.

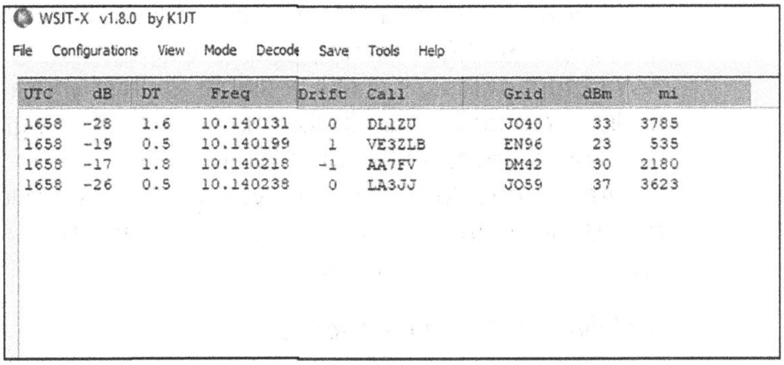

WSJT-X v1.8.0 by K1JT

File Configurations View Mode Decode Save Tools Help

UTC	dB	DT	Freq	Drift	Call	Grid	dBm	mi
1658	-28	1.6	10.140131	0	DL1ZU	JO40	33	3785
1658	-19	0.5	10.140199	1	VE3ZLB	EN96	23	535
1658	-17	1.8	10.140218	-1	AA7FV	DM42	30	2180
1658	-26	0.5	10.140238	0	LA3JJ	JO59	37	3623

Figure 5.3 — Decoded WSPR signals on 30 meters.

From left to right, the decoded text represents the following:

UTC: The time in UTC.

dB: The strength of the decoded signal.

DT: The time deviation of the received signal, in seconds. This is an indication of how closely synchronized the received station's computer clock is to your computer clock.

Freq: The RF frequency of the received signal.

Drift: The amount of frequency drift, expressed in Hertz, which occurred while the signal was being transmitted. This number can be positive or negative, depending on whether the signal was drifting up or down in frequency.

Call: The call sign of the transmitting station.

Grid: The grid square where the station is located.

dBm: The RF output power the transmitting station is using.

mi: The calculated distance between you and the transmitting station, in miles or kilometers, according to how you've set up *WSJT-X*.

As you'll see in Figure 5.3, the best received DX was a transmission from DL1ZU in Germany at 3,785 miles. You can see that he was apparently running just 33 dBm output, or 2 W.

Transmitting

To get in the game yourself, set the red marker above the waterfall display to an empty space between the other WSPR signals you see in the waterfall (if you can see any signals). Assuming that you've already set your output power level and transmit percentage, just click the ENABLE TX button. *WSJT-X* may transmit at the next available even minute, or it may not; it depends on the transmit percentage you've selected.

When it does begin transmitting, be sure to watch your transceiver's ALC metering. Even though you may be running only 5 W output or less, it is still possible to drive your rig with too much audio, creating a badly distorted, undecodable signal. It is the same precaution described in Chapter 1.

After a couple of transmit cycles, get online and go to **WSPRnet.org**. Bring up the database, or the map, and do a search of your call sign to see who heard you.

Stability: Time and Frequency

All the so-called *WSJT-X* modes require accurate time, and WSPR is no exception. Read the discussion of time in Chapter 3. In Figure 5.3, you will see that two stations are out of sync by 1.6 and 1.8 seconds. Their signals were obviously decodable, but had their time deviations exceeded 2 seconds, chances are they would not be decodable.

But was the sync issue their fault or mine? Notice that in Figure 5.3 there are also two stations that are only 0.5 seconds out of sync with the computer's time. This points the finger of fault at the two stations that are much farther out of sync. If the problem was at your end, all the stations would probably have been excessively out of sync.

Always take a moment to make sure your computer's time has

WSPR Protocol Specification

Standard message: callsign + 4-digit locator + dBm (e.g., K1ABC FN20 37)

Standard message components after lossless compression: 28 bits for callsign, 15 for locator, 7 for power level, 50 bits total.

Forward error correction (FEC): non-recursive convolutional code with constraint length K=32, rate r=1/2.

Number of binary channel symbols: nsym = (50+K–1) * 2 = 162.[2]

Keying rate: 12000/8192 = 1.4648 baud.

Modulation: continuous phase 4-FSK, tone separation 1.4648 Hz.

Occupied bandwidth: about 6 Hz

Synchronization: 162-bit pseudo-random sync vector.

Data structure: each channel symbol conveys one sync bit (LSB) and one data bit (MSB).

Duration of transmission: 162 * 8192/12000 = 110.6 s.

Transmissions nominally start one second into an even UTC minute: e.g., at hh:00:01, hh:02:01, etc.

Minimum S/N for reception: around –28 dB on the WSJT scale (2500 Hz reference bandwidth).

been recently calibrated before you begin a WSPR session. You can use software to keep your computer's time in sync (see Chapter 3). If you are using a transmit-only WSPR unit, beware that some of these can be rather "drifty" when it comes to time. A few of these models offer extremely accurate GPS time-base options and they are well worth the additional cost.

Frequency stability is also critical for WSPR. Notice that none of the decoded stations in Figure 5.3 are drifting more than plus or minus 1 Hz. This is excellent stability. A signal that drifts more than about 4 Hz is usually undecodable.

If you are transmitting WSPR with a modern transceiver, you can usually be assured of having a stable signal. On the other hand, frequency drift can be an issue if you are using a transmit-only unit. Some of these transmitters can drift due to changes in temperature or dc power. If you're thinking of investing in a transmit-only radio, make sure to look at the frequency stability specifications. If the manufacturer offers a temperature-compensated crystal oscillator (TCXO) option, it is a worthwhile purchase.

Studying Propagation

Even under poor propagation conditions, a "dead" band is usually open to somewhere. With WSPR, you'll discover that dead bands are open far more often that you may realize. For instance, let's say that you've tuned through the 10-meter band without hearing a single signal. Fire up *WSJT-X* and configure it to begin sending WSPR on the band. Check WSPRnet after a half-hour or so and you may be surprised.

Of course, this doesn't mean the band is open for casual SSB conversations along the path. It is a matter of correctly interpreting the WSPR results and applying the interpretation to other modes.

According to Dr. Joe Taylor, K1JT, "A CW signal becomes audible at signal-to-noise ratios of around –15 dB on the WSPR scale. A minimal CW QSO (such as those often made via moonbounce) is possible at about that signal level. A CW QSO becomes relatively comfortable (although signals are still weak by most standards) at S/Ns of around 0 dB.

"Another way of looking at it is that S/N = 0 dB in a 2500 Hz reference implies S/N = +10 dB in a 250 Hz bandwidth. That's enough for a good CW operator. Note that it doesn't necessarily require a 250

Hz filter in the receiver, because a skilled operator's ear-brain system can provide effective filtering down to the 50 – 100 Hz range.

"Another 10 dB, more or less, is required for reasonable copy on SSB."

With WSPR, you can watch as propagation changes at sunrise and sunset, or as it changes during other times of the day. Paths will open to parts of the world at one hour, and then close the next. It will also vary according to the time of year.

WSPR can be particularly interesting on 6 meters. You can observe a band that is closed or hours or days on end, only to see it open for a few minutes over paths of hundreds or thousands of miles.

Testing Antennas

WSPR is an excellent tool for antenna testing. You can erect a wire dipole antenna, for example, that's oriented from north to south. Test this antenna by making some WSPR transmissions and then check to the WSPRnet database to see how well your signals were received. Now re-orient the antenna from east to west and transmit again. Check the database again and you will notice a substantial difference. Your signal may be weak where previously it was strong, and vice versa.

Take care when you are evaluating the results of your tests, however. The types of antennas and radios in use at the receiving stations can vary considerably, and so will their signal reports. Someone who has a gain antenna pointed at you will report a stronger signal than, say, someone using an omnidirectional antenna.

Also, don't forget that propagation can vary rapidly at times. A station that receives your signal at –10 dB for one two-minute interval, may barely hear you at all 15 minutes later. Local noise levels can also impact how well your signal is heard.

With all this in mind, look for patterns among several reporting stations instead of focusing too much on individual reports. Look for averages rather than weak or strong exceptions. For example, do *most* of the stations south of your location receive your signal weakly, while *most* stations east of you receive you strongly? Disregard the extremes of weak or strong reports and instead try to find the average. Chances are that the average results will tell you that your antenna has a radiation pattern that favors the east over the south.

Chapter 6

VARA and *Winlink Express*

Data communication at HF frequencies has always been a challenge. Part of the problem is technical because fading, noise, and interference — or a combination of all three — can wreak havoc with data signals. It takes only a single lost bit of information to corrupt the entire communication stream and cause errors at the receiving end.

Another issue is regulatory. Due to the relative lack of spectrum on the HF bands, many nations limit the bandwidths of signals they allow at these frequencies. That places a serious constraint on how fast stations can exchange digital information since higher data rates tend to result in wider signal bandwidths.

In the United States, the Federal Communications Commission governs amateur radio with rules that are often technologically antiquated. As a result, American amateurs labor under particularly severe restrictions when it comes to digital communication at HF frequencies. For instance, the FCC's Part 97 rules state that at frequencies below the 10-meter band, digital signaling rates must not exceed 300 *baud*. (Baud defines the number of "symbols" that can be sent each second.)

In the 21st century we're more accustomed to thinking in terms of *bit rates*, especially in digital communication, because we are more interested in how many bits of information we can push through a given channel each second. There is a relationship between bit rate and baud rate, but isn't equivalent, or intuitive. You can, for example, use clever coding and various modulation schemes to squeeze many bits per second into what may seem to be an otherwise sluggish baud rate. This is how the inventors of PACTOR were able to achieve such high bit rates while still using a symbol rate of 300 baud. However, there is a limit to how much technical magic you can perform before you simply must increase the symbol rate and, eventually, the bandwidth.

PACTOR IV's signal bandwidth is less than 3000 Hz, but its symbol rate exceeds 300 baud. That puts it out of bounds for American amateurs. The ARRL has petitioned the FCC to modernize its rules in this regard, but as of early 2022 the FCC had yet to act.

So, until the rules change, hams must still work within the 300-baud chokehold — at least on the HF bands.

The Debut of *VARA*

Several software developers have continued to work on the problem of stuffing more bits through the 300-baud pipeline. At the same time, they struggle with the challenge of making such technology affordable for average amateurs.

Affordability has been an issue with PACTOR. Modern PACTOR controllers are available from only one manufacturer — Spezielle Communications Systeme, better known as SCS. Without question these are innovative devices, and they make PACTOR communication possible without requiring significant computing power. However, this innovation comes with a substantial price tag. The least expensive SCS PACTOR units were selling for more than $1,200 when this edition went to press.

The challenge, then, is to develop a less-expensive alternative to PACTOR that could rival its throughput at HF frequencies while remaining in compliance with FCC rules. Among the most successful solutions has been devised by Spanish amateur José Alberto Nieto Ros, EA5HVK, with his *VARA* software modem.

VARA is not an acronym, which is to say the individual letters don't have specific meanings. In Spanish, a *vara* is a pole or rod. In bullfighting, varas are lances used by horse-mounted picadores. Stretching the interpretation, *VARA* in this context can be considered a technological "lance" that cuts through difficult HF communication channels.

In the amateur community, *VARA* is a software modem that modulates signals for transmission and demodulates received

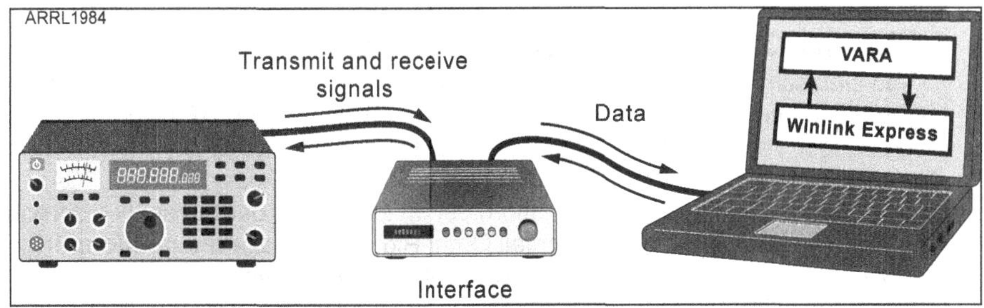

Figure 6.1 — In this example, *VARA* and *Winlink Express* are sharing information on the same computer. The computer, in turn, is connected to a transceiver interface that contains a sound device. The sound device converts audio from the radio to data for the computer, and vice versa. If you own a transceiver that already has a built-in interface, this external interface is unnecessary. In that case, the transceiver would connect directly to the computer, probably with a USB cable.

signals. Unlike PACTOR controllers, which are self-contained units requiring computers to act only as human interface devices, a software modem still needs other hardware and software to complete the communication task.

VARA software uses the computer sound device, or the sound device in your interface or transceiver, to create and decode complicated signals, keep track of errors, request repeats, and more. But it must work in conjunction with other software. As you'll soon see, in most instances that other piece of software is *Winlink Express*, an application used specifically for communicating within the Winlink HF email network. See the diagram in **Figure 6.1**.

VARA Requirements

Once you've set up a station to operate HF digital, as we discussed in Chapter 1, that same station can be used with *VARA*. That's one of the beauties of modern HF digital operating using sound-device-based software. A station used for RTTY, for example, can also operate FT8, WSPR, FT4, *JS8Call* … and *VARA*. It's just a matter of closing one application and opening another.

VARA requires the same hardware as any other mode:

- An HF SSB transceiver
- A sound device interface (if the transceiver doesn't already have one)
- A computer running Windows 10 or 11. At the time of this writing, *VARA* was not available for Linux or MacOS.

Digital Conversations on HF Using *VarAC*

Irad Deutsch, 4Z1AC, has developed a software application called *VarAC* that allows amateurs to enjoy live keyboard-to-keyboard conversations on HF using the *VARA* modem and protocol. The software is free and available for download at **www.varac-hamradio.com/download**. If you have CAT (see the article in this book's Appendix titled, "CAT — Computer Aided Transceiver" for more information) set up to control your transceiver's frequency, you may need to also download and install Omni-Rig at **www.dxatlas.com/omnirig/**.

VarAC downloads as a compressed Zip file that you extract to a folder on your hard drive. In the extracted folder look for the *VarAC* application and icon. Right click on the icon and you can send a shortcut to the Windows desktop.

If *Windows* 10's application security function blocks *VarAC* from running on your station computer, select "more information" in the blue box, and then click "Run Anyway."

The *VARA* modem should start, followed by *VarAC. Omni-Rig* will fire up in the background.

Chatting by Keyboard

VarAC is a well-designed application. There is a logging function and a cool signal strength display that shows not only your partner's signal strength, but also your signal strength as received at his end. The application can also send small files.

VarAC operators tend to congregate at several calling frequencies: 7.105, 14.105, 21.105, and 28.105 MHz. To minimize interference, the idea is to establish contact at one of these frequencies and then move elsewhere. If you are chatting on a calling frequency, after about 10 minutes you'll see a reminder to QSY. Here is where having CAT control with *Omni-Rig* is slick. You just use so-called "gesture" text to ask the other station to shift frequency, perhaps 750 Hz in either direction. QSYU sends the request to move up and QSYD invites the other station to move lower. If the other operator accepts, the frequency shift occurs automatically. If there is activity going on at the new frequency, you can QSY again.

The evaluation version of *VARA* is free for downloading at **rosmodem.wordpress.com**. Note that there is a separate version of *VARA* called *VARA* FM that is designed for use on VHF. In this book we're only addressing the HF version.

The free version of *VARA HF* allows maximum bit rates of about 180 bits per second. It also imposes a time limit when used during Winlink network connections. The paid version — selling at about $70 — allows bit rates in excess of 8490 bits per second, which rivals the performance of PACTOR III, and it removes the time limits. Once you've installed and tested *VARA*, you can use the upgrade menu option within the application to purchase a license.

VARA Setup

After download and installation, configuring *VARA* is relatively

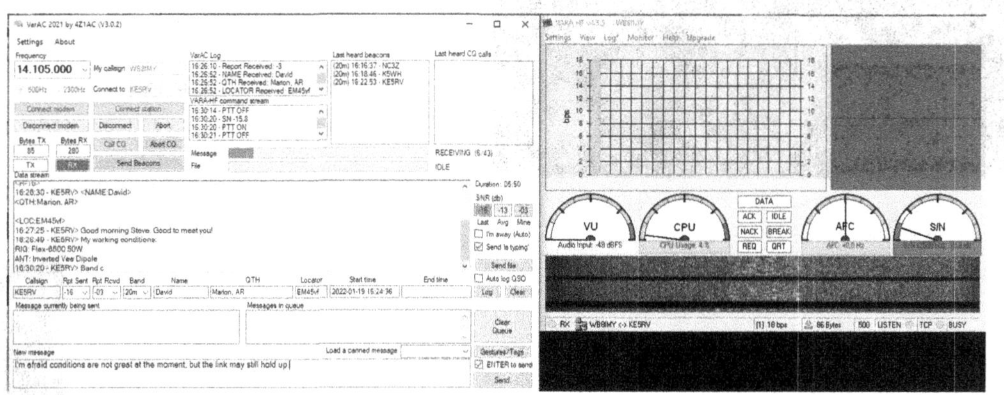

Figure 1 – Engaged in conversation with Dave Judkins, KE5RV, on 20 meters. *VarAC* occupies the left-hand window, and the *VARA* modem appears at right.

Impressions

I've enjoyed a number of *VarAC* conversations, and they remind me of similar connected chats using AMTOR or PACTOR. Once you connect with another station, it's like joining a choreographed dance; you transmit, and then the other station transmits. It all happens automatically so you need only sit back and type.

Should you have questions, there is an active Facebook group. Just search for "*VARA* operators — digital mode for HF radio."

straightforward. In fact, you can accept the default values of the application, including the *TCP* port numbers. TCP stands for Transmission Control Protocol a communications standard that enables applications and computing devices to exchange messages over a network. In the case of *VARA* and *Winlink Express*, these two ports will be used to swap information between the two applications.

Make sure your transceiver is set for upper sideband (USB), "digital," or "USB-Digital," whichever mode your radio uses for digital communication. Also check that your receive filters are set as wide as possible: 0 to 3000 Hz.

Once you've familiarized yourself with the *VARA* software, the next step is to download and install *Winlink Express*. There is no need to leave the *VARA* application running. *Winlink Express* will start *VARA* automatically when you are ready to go "live" on the air.

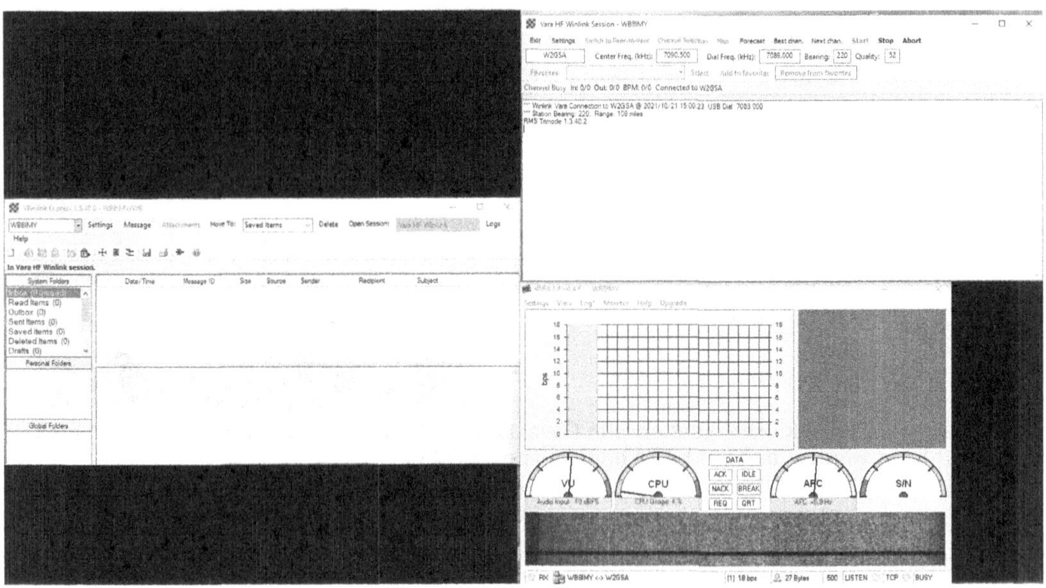

VARA attempting a connection. The _VARA_ window at the lower right includes a waterfall display and "meters" for audio input level (VU), CPU usage, automatic frequency control, and the signal-to-noise ratio (S/N).

Winlink Express

Winlink Express is the latest in a string of software applications designed to support communication with the Winlink global email network. It is the culmination of years of development by Winlink volunteers and it continues to evolve as this is being written. For that reason, we won't go into a detailed discussion of the software. By the time you read this, *Winlink Express* may have changed significantly. You'll always find the most up to date documentation at **www.winlink.org**.

Winlink Express for Windows is available free at the Winlink website (although donations are gladly accepted). When you start *Winlink Express* for the first time, you must enter your email address, call sign, and grid square location. See **Figure 6.2**. Unless you already have an account with the Winlink network, you will need to register for an account. This is free of charge.

As part of the *Winlink Express* setup, you'll need to let the software know the type of transceiver you are using and which COM ports your computer is using to communicate with it (**Figure 6.3**). You do not

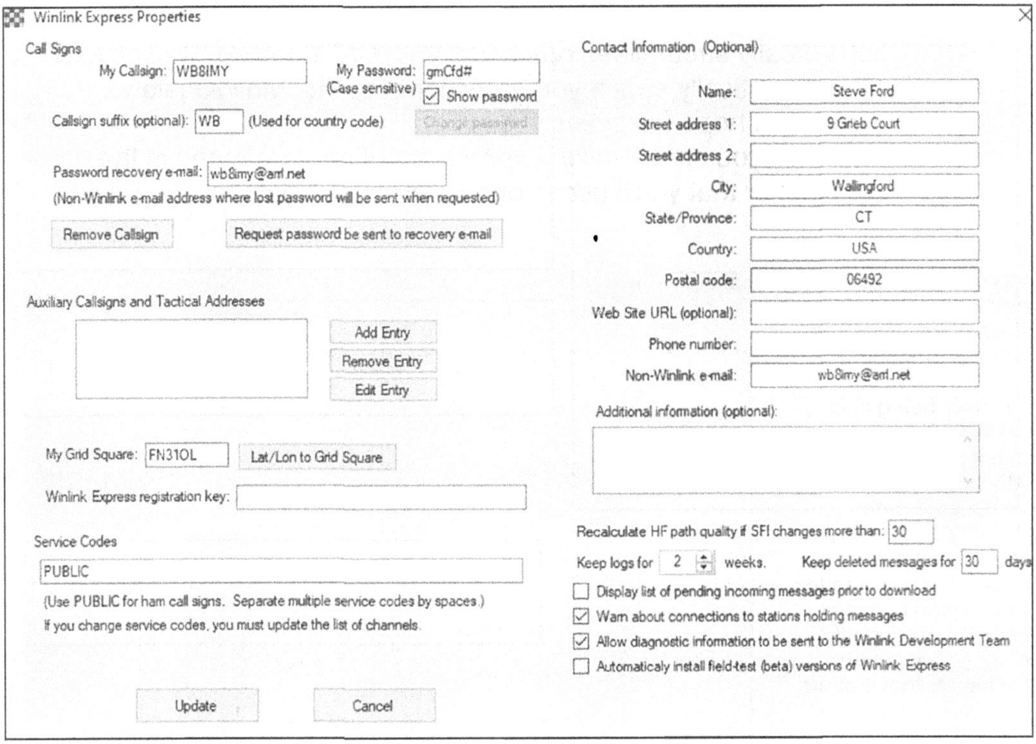

Figure 6.2 — When you start *Winlink Express* for the first time, you will be required to enter your call sign, email address, and grid square location.

Figure 6.3 — This *Winlink Express* menu allows you to specify the type of transceiver you are using and how it is connected to your computer.

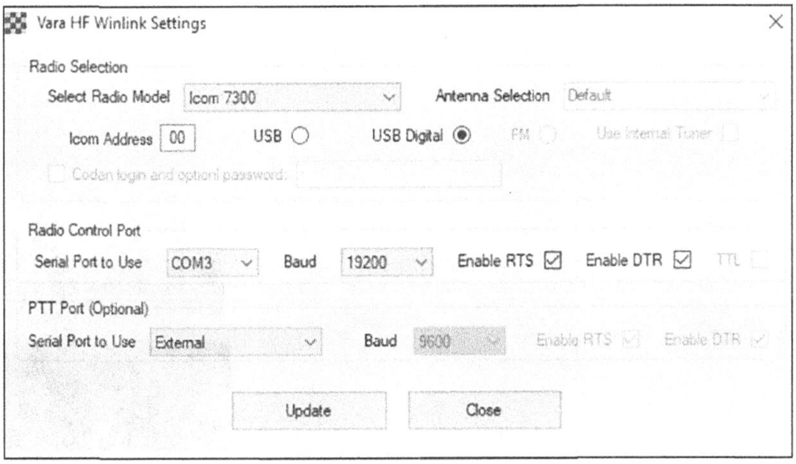

need to have your radio set up for CAT control to use *Winlink Express*, but it greatly streamlines operation. With CAT enabled, *Winlink Express* will automatically switch your transceiver to the Winlink gateway station you hope to access.

When you start *Winlink Express*, you'll notice a menu in the upper right corner that you'll use to open a "session." Use this drop-down

In the HF Channel Selector window of *Winlink Express* you'll see a list of stations you have the best chance of being able to reach. They are ranked in descending order accounting to estimated path reliability and quality. If you click your mouse cursor on a call sign in the left hand column, *Winlink Express* will set up a connection session for that station.

HF Channel Selector

Exit Select Update Via Internet Update Via Radio Map Forecast SFI All RMS

Callsign	Frequency (kHz)	Mode	Grid Square	Hours	Group	Distance (mi)	Bearing (Degrees)	Path Reliability Estimate	Path Quality Estimate
K3DO	7103.800	V2300	FM19GL	00-23	PUBLIC	281	242	80	55
NA1DX	7101.500	V2300	FM18QT	00-23	PUBLIC	274	229	80	55
N2LEE	7103.500	V2300	FM18HX	00-23	PUBLIC	297	236	81	55
N3MLB-10	7096.000	V500	FN10KJ	00-23	PUBLIC	238	253	77	54
N3MLB-10	7102.500	V2300	FN10KJ	12-23	PUBLIC	238	253	77	54
KB3PCY	7107.000	V500	FM29EV	00-23	PUBLIC	185	235	73	53
KB3PCY	3593.500	V2300	FM29EV	00-23	PUBLIC	185	235	80	53
KC2FBI	3595.500	V2300	FN21RS	00-23	PUBLIC	93	283	81	53
VE3HJL	7108.000	V2300	FN03IR	10-22	PUBLIC	365	297	76	53
VE3AWN	7107.500	V2300	FN15SM	00-23	PUBLIC	334	328	75	53
VA2XMP	7105.000	V2300	FN35BQ	00-23	PUBLIC	296	350	74	53
VA3ETN	7031.500	V2300	FN02FW	12-23	PUBLIC	359	289	76	53
W1EO	7102.500	V2300	FN42IM	00-23	PUBLIC	105	046	71	52
VA2XMP	5348.000	V2300	FN35BQ	00-23	PUBLIC	296	350	76	52
WD1O	7104.500	V2300	FN53IX	00-23	PUBLIC	247	045	72	52
VE3AWN	5373.000	V2300	FN15SM	00-23	PUBLIC	334	328	75	52
AC2SV	3591.700	V2300	FN20UB	00-23	PUBLIC	94	237	81	52

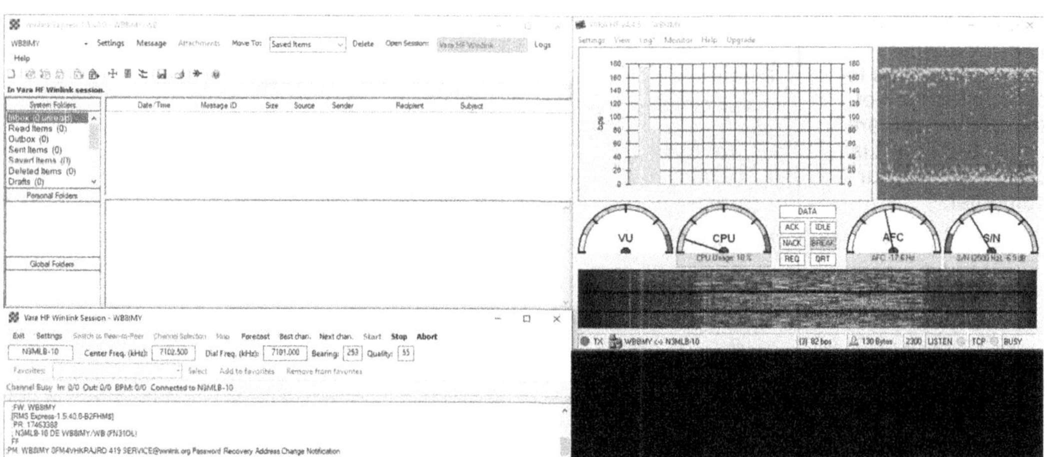

Figure 6.4 — *Winlink Express* and *VARA* have established a connection to the N3MLB-10 Winlink gateway station at 7102.5 MHz.

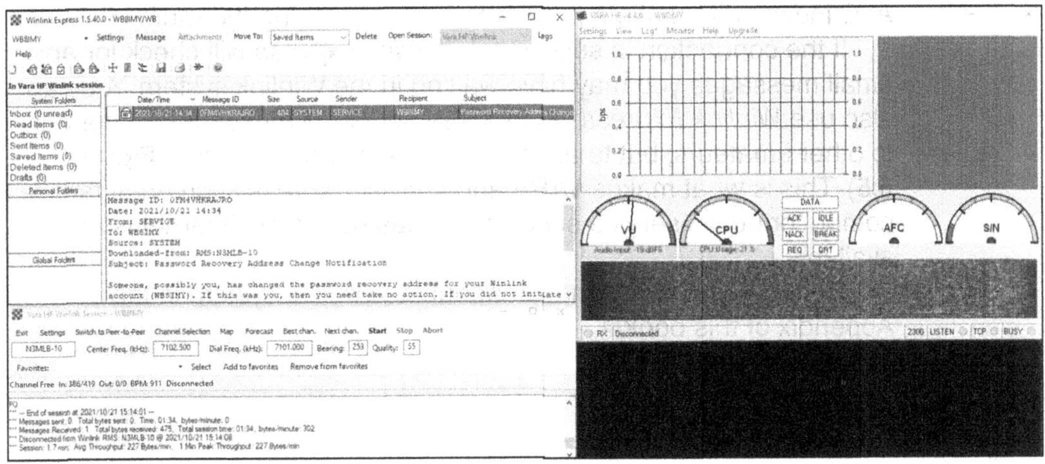

Figure 6.5 — Once the Winlink connection is established, *Winlink Express* will automatically download any waiting messages, as it has done in this example. You can also use *Winlink Express* to compose messages that can be sent to anyone with an internet connection, or a radio link to the Winlink system.

menu to select *VARA HF* as your session modem. (*Winlink Express* can also work with other software and devices, including PACTOR controllers.) When you open a session in this fashion, *Winlink Express* will start *VARA* automatically and you'll see its screen appear on your monitor, along with several indicators and a waterfall display. Make sure the *VARA* VU "meter" is not indicating in the red; that means the receive audio level is too high.

In the smaller *VARA HF Winlink Session* window, click your mouse cursor on CHANNEL SELECTION. A new window opens that shows you all the available Winlink Gateway stations. If your computer is connected to the internet, *Winlink Express* will attempt to access the latest propagation data and use it to calculate which stations are likely to be accessible from your location.

Select a station by double clicking on its call sign and it will be added to the *VARA HF Winlink Session* window. If your transceiver has CAT enabled, Winlink Express will tune your radio to the correct frequency automatically. If not, you'll need to put the radio on frequency manually.

In the session window, click START and *VARA* will go to work, attempting to make the connection. See **Figure 6.4**. As with other sound-device-based modes, keep a careful eye on your transceiver's

ALC indicator to make sure you are not overdriving the radio.

If the connection is successful, *Winlink Express* will check for any email messages you may have waiting in the Winlink system. You can also use *Winlink Express* to compose and send messages — not only to other amateurs, but to anyone who has internet access (**Figure 6.5**). This is what makes *VARA* and *Winlink Express* such an attractive combo for public service support in areas where the internet is not available.

You'll find more information about the Winlink system in the Appendix of this book.

Chapter 7

JS8 and *JS8Call*

The FT8 digital mode was wildly popular when this edition went to press. In fact, it is accurate to say that FT8 is the most widely used HF digital communication mode in amateur radio today. It's cousin, FT4, is not far behind. We discussed both modes in Chapter 3.

For many amateurs, however, the main shortcoming of both FT8 and FT4 is that you can't enjoy a true conversation. A contact made with FT8 or FT4 is just that — a "contact" only. The contact is valid for contests and awards, and it provides a highly objective signal report, but a conversation between stations is impossible.

Several years ago, Jorden Sherer, KN4CRD, wondered if it might be possible to combine the robust ability of FT8 to communicate at weak signal levels with a conversational component. The *WSJT-X* software suite, which includes FT8 and FT4, is open source and its code is available to experimenters, so Jordon began using it to explore the possibilities.

The result was a modified form of the FT8 that Jordan called JS8. He then constructed a directed-calling messaging protocol using JS8 and brought it to life in a software application known as *JS8Call*. Many amateurs confuse the mode — JS8 — with the software, *JS8Call*. In this chapter we'll discuss the software.

JS8Call is free and available for computers running *Windows*, *Linux*, and *MacOS*. You can download the files at **files.js8call.com/latest.html**.

Setup

If you've already installed *WSJT-X* at your station, setting up *JS8Call* will be familiar (**Figure 7.1**). You'll need to enter your call sign and grid square designator. And as with *WSJT-X*, if you are using an external interface to connect your computer to your transceiver, you will need to specify the COM ports for keying and CAT control (if used), as well as the audio input and output devices. See the *WSJT-X* setup discussion in Chapter 3.

Timing is critical in JS8, just like it is with FT8 and FT4. You must synchronize your computer time to an accurate external source. This is described in Chapter 3 as well.

Your transceiver must be set to USB, "digital," or USB-D (consult your transceiver manual). Beware of overdriving the audio input and causing excessive ALC activity.

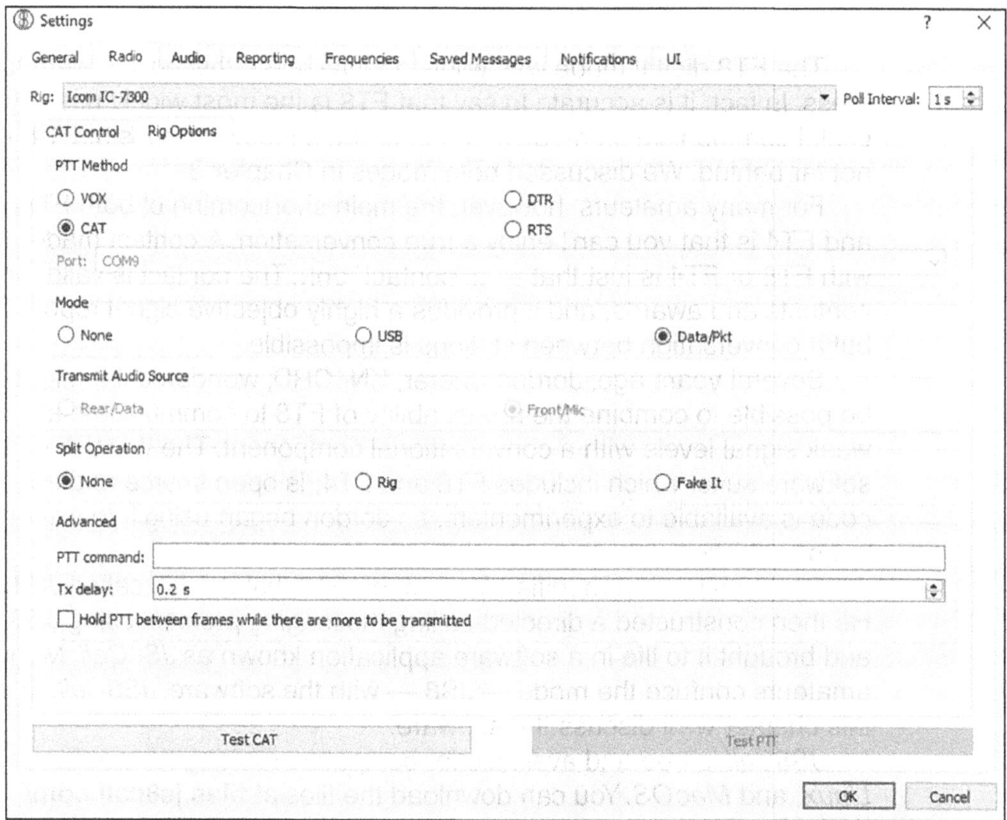

Figure 7.1 — The Settings menus in *JS8Call*. You'll find it under File in the upper-left corner of the screen.

On the Air with *JS8Call*

If you've operated FT8, you know that each 12.6-second transmission sends only 13 characters (including spaces). The same is true for *JS8Call*, but the software chops your message text into 13-character segments and sends each segment individually. So, at the receiving end, the full text is reassembled, one segment at a time. Yes, this makes for a slow conversation, but keep in mind that you're also enjoying a level of sensitivity similar to FT8. For amateurs with compromise antennas and low RF output levels, this is a major advantage.

As we do elsewhere in this book, it is important to caution the reader that software is ever changing, especially software that is under active development like *JS8Call*. For this reason, the discussion of *JS8Call* in this chapter will be limited to those aspects that are least likely to change in the near future. For the latest information, see Jordan's excellent guide at **files.js8call.com/latest.html**. Much of the information in this chapter is based directly on Jordan's *JS8Call* documentation.

JS8Call Speeds

At the time of this writing, *JS8Call* could operate in four different speed modes. As you'll see below, slower modes permit decoding at weaker signal levels.

Mode	*Frame lengths (seconds)*	*Bandwidth (Hz)*	*Speed (WPM)*	*Minimum Decodable Signal Level*
Slow	30	25	8	–28 dB
Normal	15	50	16	–24 dB
Fast	10	80	24	–20 dB
Turbo	6	160	40	–18 dB

The idea is to start your contacts in the Normal mode and then shift to Fast or even Turbo if conditions permit. Or, if conditions become difficult, downshift to the "slow" mode.

Navigating the JS8Call Screen

Look at the *JS8Call* example in **Figure 7.2**. Band activity is displayed in the large "Messages" window at the left. Call activity (call signs you've heard) appear in a column on the far right. Right clicking on a call sign in the call activity window will show a menu with an

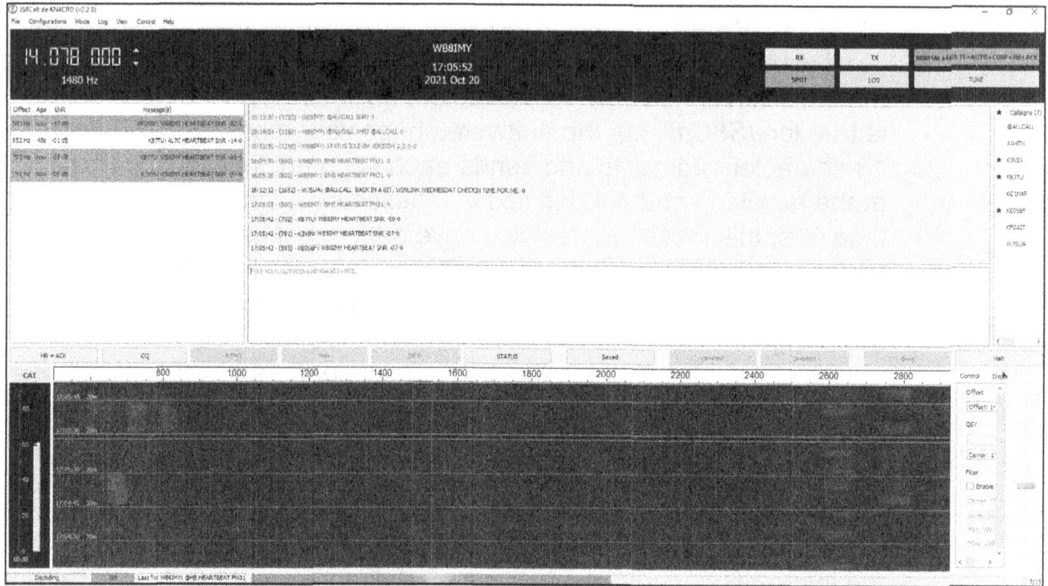

Figure 7.2 — When you start *JS8Call*, the full screen appears on your monitor. In this example, CAT control was enabled with the transceiver, so the frequency appears automatically in the upper-left area. If you have CAT control, you can change from one JS8 frequency to another by just right clicking your mouse cursor on the frequency display and selecting your new "destination" from the drop-down menu.

option to move your transmit/receive offset to that audio frequency and send specific messages.

The waterfall portion of the screen displays the signals that are presently within your audio passband. You can click on the waterfall to set your audio frequency offset.

There is also an option to change your VFO frequency to center your selected audio offset to the rig passband center. This allows you to use narrow filters easily and is helpful for rigs with non-linear passbands.

By opening the waterfall controls (Click VIEW; then select SHOW WATERFALL CONTROLS) you can configure your waterfall display, access a filtering feature (limiting which frequencies the decoder will try to decode), and the timing feature (allowing you to drift your local time sync to match a station — a feature not found in *WSJT-X*).

The top center-right box, shaded in gray, displays messages that are either on the frequency offset you're on, or that come from stations that have directed a message to you (they sent a message that included your call sign). Directly below this box is the area in which you type your messages for transmission.

As you type your message, you'll notice that the SEND button displays the transmission time it'll take to send your complete message. To start sending, you need only click the SEND button. *JS8Call* will begin sending your message, one frame at a time (see **Figure 7.3**). With each transmission the button will update the remaining time. *JS8Call* supports typeahead, so you can start transmitting and continue typing your message as each frame is transmitted.

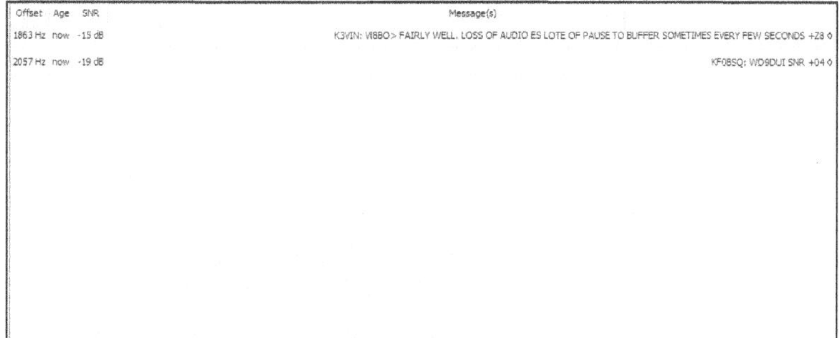

Figure 7.3 — Here is a portion of the Message(s) window showing the text of an actual contact taking place between K3VIN and W8BO. K3VIN's text was split into 13-character segments and each was sent separately. As each segment was received, it was appended to the one before to create a coherent stream of text.

Message Types

There are three primary *JS8Call* message types: Standard, Undirected, and Directed.

Standard messages are free-text messages that do not start with a call sign or a directed command. These messages will only print at other station locations if they align their receive frequency within 10 Hz of your transmit frequency. This operation is similar to other keyboard-to-keyboard digital modes like Olivia, RTTY, and PSK.

Undirected messages are exactly what the term implies. They can be read by anyone and do not have to be sent at a particular frequency.

Directed messages are special transmissions that automatically prefix your message with your call sign. Directed messages are useful because you do not have to include your call sign in your message, allowing you to use more of the transmission frame(s) for the actual message text. It also alerts the recipient that a message was sent to them. As long as you are in the same passband, you do not have to be on the same frequency to receive a directed message.

To send a directed message, all you need to do is include the call sign of the receiving station as the first word in the message or select a call sign in your heard list (in the column at the far right) to have it automatically prefixed.

Group Messages

Group messages go to a group destination rather than to an individual station. For example, there is a special **@ALLCALL** group call sign that you can use to send the message to anybody who is able to receive your message. If you send a message like this . . .

@ALLCALL CQ CQ CQ

...it will be seen by everyone who can decode the message.

Group call signs always begin with @ character and can be up to 8 alpha-numeric (A-Z 0-9) characters in length.

Group call sign functionality allows you to direct your message to anybody who has "joined" the group. You join the group by adding the group name to your settings. All stations who want to receive group messages must add the group to their station configuration. Stations without the group will still be able to see the message received in the band activity, but those messages will not be directed to them.

If you wanted, for example, to send a message to the Ohio ARES group and your call sign was WB8IMY, you'd send . . .

@ARESOH QSL?

You don't have to type in your call sign — just **@ARESOH QSL?** This group call sign will behave similarly to @ALLCALL. Everybody who has added the @ARESOH group to their station configuration will have the message printed on their screens. If instead, you transmitted:

@ARESOH SNR?

All group member stations who have AUTO enabled in their *JS8Call* setups would respond with a signal report, as if you had queried each group station individually.

JS8Call has a number of built-in group call signs. See Table 7.1.

There are also two "special" groups for spotting. When spotting stations received messages sent to these @JS8NET and @APRSIS groups, the messages are posted to the JS8NET spotting server for processing. This allows for specialized functionality to be built to

Table 7.1
Group Call Signs Included in *JS8Call*

Continental DX Groups	Generic Groups	Operator Groups	@EMCOMM	@QSOPARTY
@DX/NA	@GROUP/0	@COMMAND	@ARES	@CONTEST
@DX/SA	@GROUP/1	@CONTROL	@MARS	@FIELDDAY
@DX/EU	@GROUP/2	@NTS	@AMRRON	@SOTA
@DX/AS	@GROUP/3	@NET	@RACES	@IOTA
@DX/AF	@GROUP/4		@RAYNET	@POTA
@DX/OC	@GROUP/5	Special Groups	@RADAR	@QRP
@DX/AN	@GROUP/6	@JS8NET	@SKYWARN	@QRO
	@GROUP/7	@APRSIS	@CQ	@JS8NET
ITU Regions	@GROUP/8	@RAGCHEW	@HB	@APRSIS
@REGION/1	@GROUP/9	@JS8	@QSO	
@REGION/2				
@REGION/3				

handle these messages.

The @APRSIS group is an experimental feature allowing APRS messages to be spotted to the APRS-IS gateway. Two message commands are available, GRID for spotting your callsign at a specific location and CMD for sending a raw APRS packet.

For example, any station receiving this message:

WB8IMY: @APRSIS GRID FN04TV53

Will submit that spot to JS8NET and spot the call sign at that location to the APRS network. You would then be able to query that spot in an APRS client, like **https://aprs.fi**.

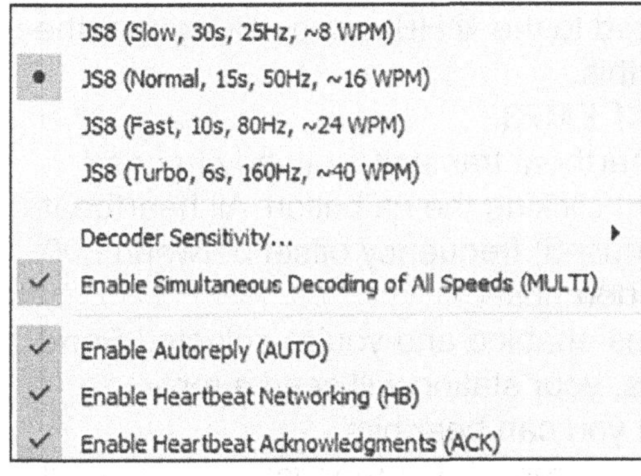

Figure 7.4 — In the Mode dropdown you can enable automatic replies, as well as Heartbeat networking and acknowledgements.

Special Directed Messages and Automatic Replies

When setting up *JS8Call*, it is a good idea to enable the AUTO (Autoreply) function so your station can send automatic replies. You'll find this in the Mode menu that's accessible at the top left area of the *JS8Call* screen (see **Figure 7.4**). An automatic reply is triggered whenever you receive a special directed message (such as INFO?). You won't have to lift a finger to respond. Of course, you can send such messages and receive automatic replies as well.

These special directed messages consist of a call sign followed by one of the following commands:

SNR? - What is my SNR?

GRID? - What is your grid locator?

INFO? - What is your station information?

STATUS? - What is your station status message?

HEARING? - What stations are you HEARING?

For example, a message could be sent to WB8IMY requesting that it reply with its station information message:

WB8IMY INFO?

If WB8IMY decodes your message, it would respond with something like this: **100 W to an end-fed long-wire antenna in central Connecticut, USA**

You can set up your station information message in the *JS8Call* configuration menus and change it at any time.

Heartbeat Transmissions

The so-called "Heartbeat" is another type of message that triggers an automatic response, but with a special twist. Simply put, it is a clever way to keep track of who you can communicate with at any given time.

You can turn on the heartbeat HB button by selecting ENABLE HEARTBEAT NETWORKING from the MODE menu as shown in Figure 7.4. An HB (or HB+ACK) button will then appear on the left-hand side of the strip of buttons in the middle of the *JS8Call* screen. If you click your mouse cursor on this button, *JS8Call* will transmit your grid square to the heartbeat network (directed to the @HB group call sign) at the next available opportunity. Like this:

WB8IMY: @HB HEARTBEAT EM73

This interval at which the heartbeat transmits can be changed from the control menu or by right clicking the HB button. All heartbeats are transmitted on a random (unused) frequency offset between 500 and 1000Hz to help reduce interference.

When you have AUTO replies enabled and you've selected **Send Heartbeat Acknowledgements**, your station will send a reply to signal to the other operator that you can hear him.

The intent of heartbeat is not to report on propagation. Instead, it is to help populate your call activity (the heard list on the right side of the screen) so you know who's likely to be reachable to make contact.

Heartbeats are not designed to start conversations. When you turn Heartbeat on, you're "joining" the heartbeat network. This network allows for planning of relays and sending messages to be stored at those receiving stations.

Station Log

JS8Call provides a handy station log. There's a log item in the main menu of the application, but you can also press the F5 key to start a log entry. The software will do its best effort to pre-populate log fields. However, you'll have to fill out some missing information manually.

The log is stored in *JS8Call*.log and *JS8Call*.adif in the log directory (which you can find by clicking on the FILE dropdown and selecting OPEN LOG DIRECTORY in the main menu).

Currently, the logging function in *JS8Call* will log each contact, according to the ADIF spec, as MFSK mode and JS8 sub mode. There is also an option in the Logging settings to log the mode as DATA instead of MFSK and JS8.

Calling CQ with *JS8Call*

One of the best ways to start a conversation is to call CQ. You'll find JS8 activity on many bands, but the most common frequencies are listed in **Table 7.2**.

In *JS8Call* the default way to call CQ is to use the "CQ CQ CQ" message. You can trigger a CQ transmission by simply clicking your mouse cursor on the CQ button on the button strip. The CQ button is usually located just to the right of the HB button. Your CQ will be sent to the @ALLCALL group so that everyone will have a chance to see it. Your grid square designator will be included as well. See the CQ example in **Figure 7.5**.

You can also send CQs at intervals by right clicking the CQ button and selecting a repeat interval. This will cause your station to repeat your CQ transmission until a reply is received.

Table 7.2
JS8 Frequencies

Band (meters)	*Frequency (MHz, USB)*
160	1.842
80	3.578
40	7.078
30	10.130
20	14.078
17	18.104
15	21.078
12	24.922
10	28.078

Replying to a CQ

The default way to reply to a CQ is to double click your mouse on the CQ message in the Message(s) window

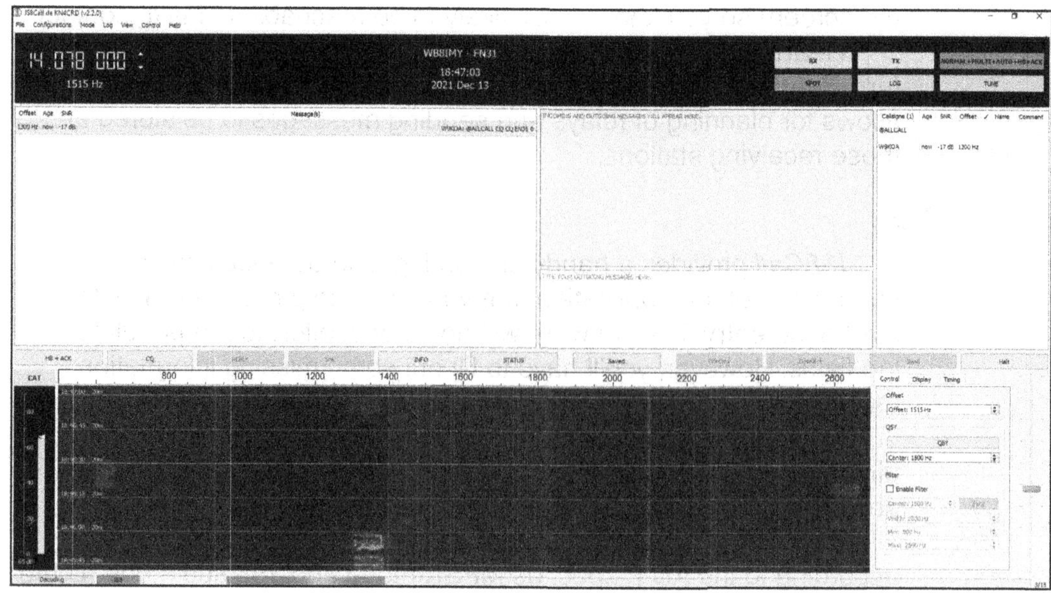

Figure 7.5 — In the Message(s) window you can see that W9KOA is sending a JS8 CQ call directed to @ALLCALL. Since the CQ was directed to @ALLCALL, any station capable of decoding the signal will see the CQ.

to highlight it. Now type the message HW CPY? and click the SEND button.

This allows the caller to choose who to connect with (he may get several replies at once). Here's an example exchange:

WB8IMY: @ALLCALL CQ CQ CQ FN31

DR4CNK: WB8IMY HW CPY?

WB8IMY: DR4CNK SNR -12 TU FOR CALL QSL?

DR4CNK: WB8IMY RR -10 FB INTO GO28. BANDS IN GREAT SHAPE

And the conversation goes on from there!

Chapter 8

PSK, Olivia, Contestia, and MFSK

Much has changed in the HF digital world in the last 20 years. At the turn of the century, hams were just beginning to use computer sound devices as digital modems. That's when the first weak-signal sound-device mode burst onto the scene: PSK31.

PSK31 would quickly become the most widely used mode for HF digital communication. This so-called "sound card revolution" would spawn several other modes in the years to follow. Before the first decade of the 21st century was over, Dr Joe Taylor, K1JT, would unleash JT65 — a mode not originally intended for HF use, but one that would rapidly climb to the top spot. JT65 offered stunning weak-signal performance that made it possible for hams to make DX contacts even with low power and minimalist antennas.

JT65's dominance ended in 2017 when Dr Taylor announced FT8 — an HF digital mode that was much faster than JT65 while still offering the same performance. In less than a year JT65 had all but vanished and FT8 was crowned the new king of HF digital. It remains so to this day.

So, what happened to all the "old" sound-device modes? Were they totally abandoned?

A few modes became essentially extinct as their presence on the air dwindled to nearly zero. Old king JT65 is rarely heard today and its once-popular cousin, JT9, has vanished.

Several other modes remain in use, albeit infrequently. Each of these "legacy" modes has its small group of fans who still enjoy putting their signals on the air. Part of the attraction of these modes is that they allow genuine conversations, something that FT8 does not. However, JS8 and the *JS8Call* software, which we discussed in Chapter 7, offer similar performance compared to FT8, but include the ability to carry on free-form conversations. *JS8Call*'s growing popularity is diminishing activity among the legacy modes, but several continue to be heard from time to time. In this chapter we'll briefly examine a few of these modes that you can still find on the air today.

Multimode Software

We discussed multimode mode software in Chapter 1, but the topic bears repeating here. It is thanks to multimode software that amateurs are still able to enjoy legacy modes today. These applications support an impressive array of modes — not only the ones we're about to profile, but also several you may have never heard on the air. If you are curious about trying some of these off-the-beaten-path modes, it is worthwhile to invest in a multimode application for your station. (Dr Taylor's *WSJT-X* software is also multimode, but it only includes FT8, FT4, WSPR, and other modes he and his team have developed.)

Here is a brief list of multimode applications to consider:

Windows

- *MixW 4* **rigexpert.com/products/software/mixw-4/**
- *MultiPSK* **f6cte.free.fr/index_anglais.htm**
- *Fldigi* www.w1hkj.com
- *Ham Radio Deluxe* **www.hamradiodeluxe.com**

MacOS

- *Cocoamodem* **www.w7ay.net/site/Applications/cocoaModem/**
- *Multimode* **www.blackcatsystems.com/software/multimode.html**

Linux

- *Fldigi* **www.w1hkj.com**

With a single multimode application, you can indulge in any digital mode you wish. The hardware requirements are the same from one mode to another, so it is just a matter of selecting the one you want from the application's menu.

RS ID

When choosing multimode software, try to find an application that supports Reed-Solomon Identification, better known as RSID. RSID is a digital signal that identifies the type of mode in use. You'll hear it as a brief warbling sound just before the main transmission. If you have RSID enabled in your software, you will see a notice appearing on your screen. Click on the box and your software will automatically select the matching mode and will set your transmit/receive audio frequency accordingly.

RSID is an excellent tool to use when you are trying to scare up a conversation on a legacy mode. It will "advertise" your presence to any other multimode software users who can receive your signal and will make it easy for them to join you.

Frequent use of RSID is strongly recommended, with one exception: PSK31. As you're about to see, PSK31 conversations take place around known frequencies, so any signals you hear on those frequencies are likely to be PSK31; there is no need to use RSID to flag them. An exception might be variants of PSK such as PSK125. This is an uncommon mode to hear at PSK "watering hole" frequencies, so sending an RSID signal may help you make contacts.

In most multimode software applications that include RSID you must

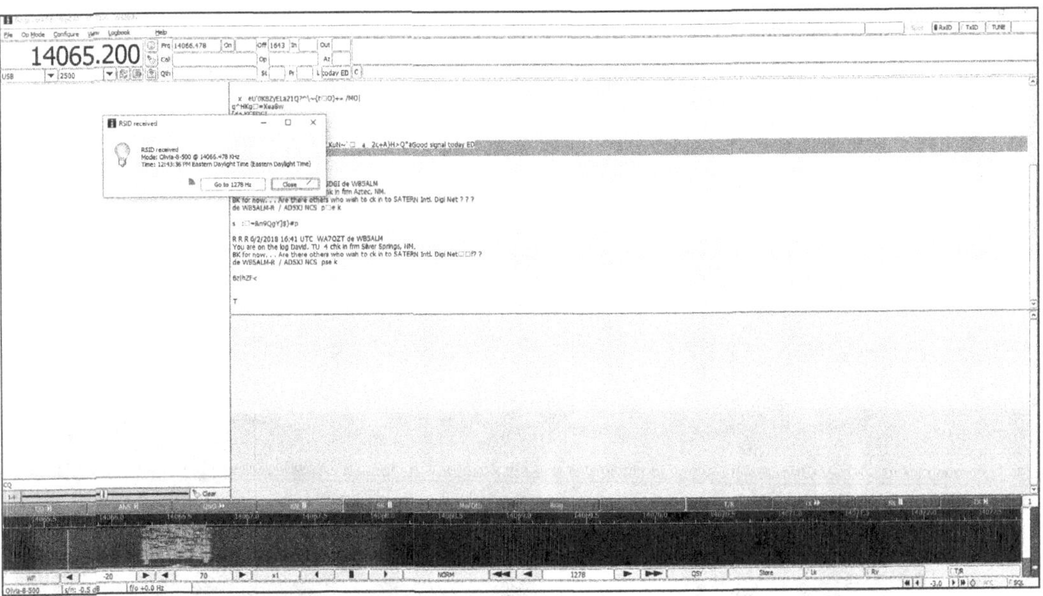

Here's an example of RSID in action. The station has decoded an RSID notification that the transmission about to take place will be using a form of Olivia (notice the RSID window that has appeared in the upper-left area of this image). If you click on the "Go to" button within the notification box, the software will configure itself accordingly.

enable reception, transmission, or both. RSID isn't the default, so you have to turn it on — usually by clicking your mouse cursor on a button.

Phase Shift Keying (PSK)

Table 8.1
Popular PSK31 frequencies

3580 kHz
7070 kHz
10140 kHz
14070 kHz
21070 kHz
28120 kHz

Phase-Shift-Keying signals, or PSK, come in several varieties, but the most common by far is PSK31. Visit the PSK frequencies (see **Table 8.1**) and you'll usually hear several signals, especially on weekends.

PSK signals will appear as vertical lines on your software waterfall display. PSK31 signals create lines that resemble railroad tracks. Click your mouse cursor on any of these lines and you should begin to see text on your screen.

You may notice that operators are occasionally using

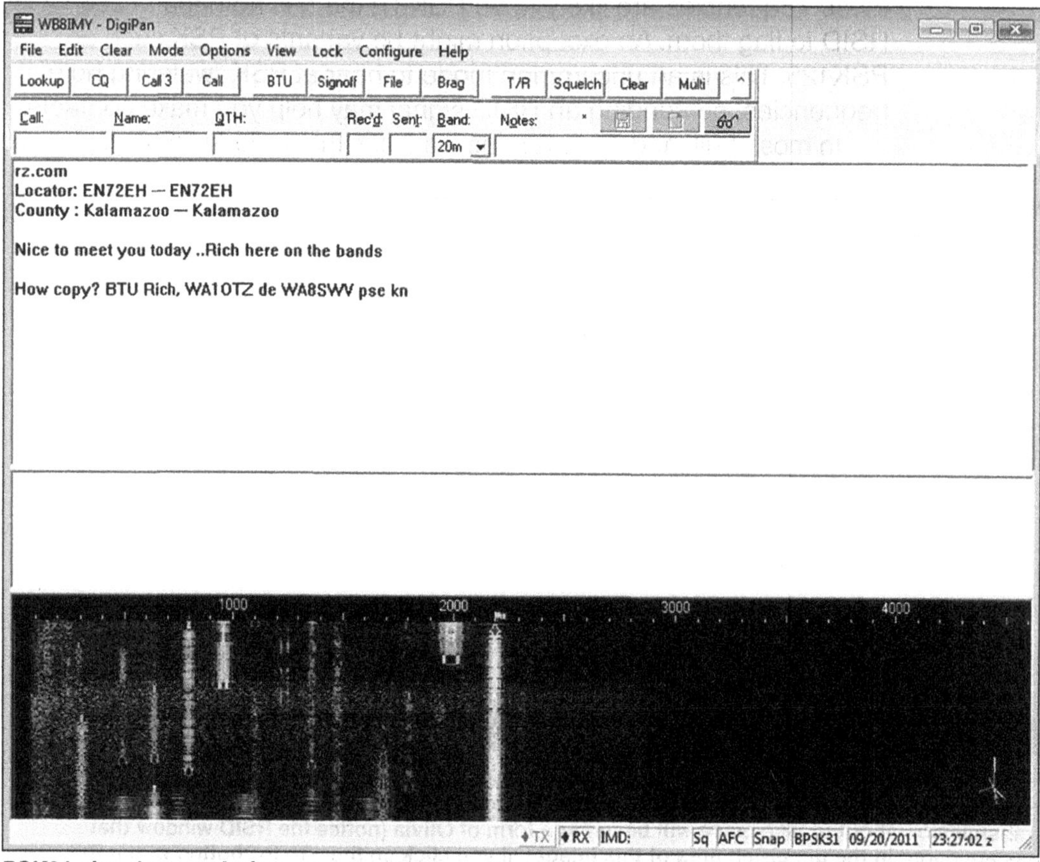

PSK31 signals seen during a contest. Each vertical line represents an individual signal.

macros — preformatted text that is stored in their software and sent with the click on a virtual button. The most common macro is the one used to call CQ. When properly configured, it will send several lines of CQ along with your call sign, and then return your radio to the receive mode. Every software application takes a different approach to creating macros, but you can find instructions in the software "help" sections.

Be careful about relying too much on macros. They are convenient — especially the CQ macro — but they can also cause confusion if used improperly. In addition, some hams create macros that contain lengthy descriptions of their station equipment, families, etc. It's tempting to tap that macro key and sit back while your computer sends more information than the other person really wants to know!

Finally, many amateurs complain — and with some justification — that overreliance on macros takes the pleasure out of making contacts. A contact based on macros is little more than a brief exchange of canned text. A ham who is expecting a leisurely chat will be gravely disappointed if all you do is send a few macros instead.

If you'd like to commune with other PSK enthusiasts, one worthwhile destination online is the PODXS 070 Club at **www.podxs070.com**. This is a group of active operators that sponsors awards as well as several on-air contests throughout the year.

Olivia

Olivia was the brainchild of Pawel Jalocha, SP9VRC, and it was named after his daughter. It is one of the better conversational digital modes in terms of its ability to exchange readable text in poor conditions. The signal can be decoded even when it is 10 to 14 dB below the noise floor (i.e., when the amplitude of the noise is slightly over 3 times that of the signal). It can also decode well under noise, fading, and interference.

Olivia has many formats, some of which are considered standard, and they all have distinctive characteristics. The formats vary in bandwidth (125, 250, 500, 1000, and 2000 Hz) and number of tones used (2, 4, 8, 16, 32, 64, 128 or 256). This makes it possible to have *40* different Olivia formats which have different characteristics, speeds, and capabilities. Luckily only a few are commonly used.

You'll find Olivia in multimode software such as *MixW*, *MultiPSK*, *Fldigi* and *Ham Radio Deluxe*. At the time this book went to press, Olivia was *not* offered in the two most popular Mac OS programs:

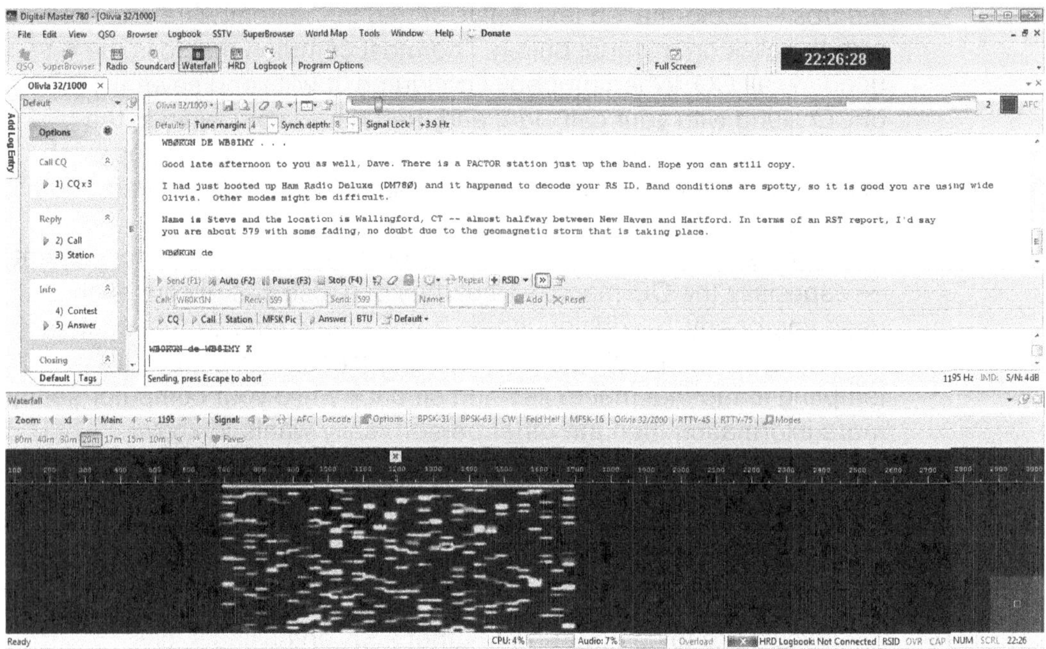

A conversation taking place with Olivia 32/1000.

Cocoamodem and *Multimode*.

The standard Olivia formats (bandwidth/tones) are 125/4, 250/8, 500/16, 1000/32, and 2000/64. However, the most used formats in order of use are 500/16, 500/8, 1000/32, 250/8 and 1000/16. After getting used to the sound and look of Olivia in the waterfall, though, it becomes easier to identify the format when you encounter it. About 98% of all current Olivia activity on the air is one of the 7 following configurations: 1000/32, 1000/16, 500/16, 500/8, 250/8, 250/4 and 125/4.

Common Olivia frequencies are shown in **Table 8.2**. Olivia announces itself with a musical sound not unlike MFSK16, but it will appear to be wider in the software waterfall display.

On the Air with Olivia

Place your transceiver in the Upper Sideband (USB) mode and start tuning around. If you suspect you've found an Olivia signal, observe its appearance in your waterfall display. Choose an Olivia mode from your software MODE menu that results in a tuning bar (or bracket) that appears to match the width of the Olivia signal in the waterfall. Position the bar on the signal and wait patiently.

If you don't see coherent text after 15 seconds or so, you may have chosen the right bandwidth but the wrong number of tones. Quickly go back to the MODE menu and try again with a different number of tones in the same bandwidth.

An experienced Olivia operator knows that this initial decoding process can take a while, so it's a good idea to send a very long CQ. An Olivia CQ lasting more than a minute is common. The idea is to give listeners plenty of time to lock in the signal.

If your software supports it, try creating an "Auto CQ" macro. This macro sends a long CQ, pauses for about a minute then sends again. The Auto CQ function will allow you to putter about your station while you wait for someone to tune you in and answer. And as mentioned earlier in this chapter, be sure to enable RSID if your software supports it.

Once you've established a conversation, sit back, relax, and enjoy. Olivia moves at a slow pace, printing only a few words on your screen at a time. This results in conversations that may last as much as an hour, perhaps even longer. Just watch the text in the receive window and type your responses in the transmit window. You don't have to wait for the other station to stop before typing your responses. On the contrary, you can enter your comments while the other station is still transmitting. When you enter the transmit mode, they'll be sent automatically.

Olivia Hints

There are a few things you can do to improve how well Olivia works at your station. Some of these tips may not necessarily be supported in the software you are using, but chances are that most will.

• **If your multimode application has a software squelch function, set it to "low" or "off."** Yes, this setting will produce some random "garbage" text on your screen, but it will also allow you to see very weak Olivia signals that would otherwise be invisible.

• **Exercise CQ patience.** We've touched on this already, but it bears repeating. When you call CQ on this mode be patient and wait at least 45 – 60 seconds before you call again. Remember that the listening station may be effectively 5 to 20 seconds behind you, still decoding your signal after you've stopped transmitting. If you instantly hit the CQ macro button again, the other person may still be decoding and won't be able to respond. Wait so that the person receiving has plenty of time to decode all your text and begin transmitting a reply.

• **Don't get carried away with your Sync Integration Period.** This software setting (some programs may label it differently) determines how "deep" the software digs into the data to decode an Olivia. You may say, "Set it as high as possible for maximum performance," but this would be a mistake. If you set this parameter sky high, you may be waiting *minutes* before Olivia begins decoding. This is clearly absurd. If you are in doubt, go with whatever the software default setting happens to be. Feel free to experiment, though. You'll eventually find that you can indeed squeeze out better performance at higher settings. It is just a matter of how much delay you care to tolerate.

• **Keep your "Search" setting at about 8.** The Search setting (or Tune Margin) determines how far from the center frequency (in the middle of the tuning bar) the software will go to grab the Olivia signal. Once again, if you are in doubt, use the default value.

Contestia

Contestia is a digital mode that is directly derived from Olivia, although it is not as robust. It is more of a compromise between performance and speed. It sounds and looks almost identical to Olivia, and can be configured in as many ways, but has just over twice the speed.

The Contestia mode has 40 formats just like Olivia — some of which are considered standard, and they all have different characteristics. The formats have a variation in bandwidth (125, 250, 500, 1000, and 2000 Hz) and number of tones used (2, 4, 8, 16, 32, 64, 128, or 256).

The standard Contestia formats (bandwidth/tones) are 125/4, 250/8, 500/16, 1000/32, and 2000/64. The most used formats are 250/8, 500/16, and 1000/32. The increased speed of Contestia is achieved by using a smaller symbol block size (32) compared to Olivia (64) and by using a 6-bit decimal character set rather than the 7-bit ASCII set that Olivia uses.

This reduced character set does not print out in both upper and lower case (like RTTY in that regard). If you're attempting to use Contestia in a public service application, this factor can be problematic for some traffic nets.

In terms of on-air activity, Olivia is the more common of the two,

Table 8.2
Common Olivia Frequencies
All frequencies are Upper Sideband (USB).

Band (Meters)	*Frequency (kHz)*
160	1835 to 1838
80	3583.25 and 3577
40	7035 to 7038
30	10141 to 10144
20	14.072 to 14075.65 and 14106.5
17	18102.65
15, 12, 10	1 – 2 kHz above PSK31 activity: 21.072, 24.922, 28.122

but it isn't unusual to pick up a Contestia CQ on occasion. Some amateurs enjoy the enhanced speed Contestia provides and don't mind the slight performance tradeoff.

You'll find Contestia signals at or near the Olivia frequencies shown in Table 8.2.

MFSK

MFSK — Multi-Frequency Shift Keying — is really a type of super-RTTY. Instead of using just two tones as RTTY does, MFSK uses many more. Sixteen-tone MFSK (MFSK16) is the most common.

The MFSK technique was developed during the heyday of teleprinter HF communications to combat multipath propagation problems, providing reliable point-to-point communications with relatively simple equipment. Piccolo, for example, was a similar mode used on diplomatic links between England and Singapore, and it typically provided good copy for an hour after the RTTY links had faded out. The technology was electromechanical then, but several important principles were recognized at the time:

- The performance (reduced error rate) improved as the number of tones used increased.
- The performance was best when the least number of symbols was used to represent each transmitted text element.
- With a special integrating detector, tones as closely spaced as the baud rate could be uniquely detected without crosstalk.

Piccolo used two symbols per text character — compare this with 7.5 symbols per character for RTTY and anywhere from three to twelve for PSK31. MFSK16 uses only *one* symbol per signaling element. With MFSK modes the baud rate (rate at which symbols are transmitted) is rather lower than the text rate. This is because each symbol carries more information in its frequency properties than RTTY or PSK. While this may seem confusing, the technique has the advantage that the longer symbols are easier to detect in noise, have a lower bandwidth, and are much less affected by multipath errors.

Piccolo originally used as many as 32 tones, but the most common form uses 6. MFSK has been tested with as many as 64 tones; the weak signal variant MFSK8, uses 32.

Anatomy of an MFSK16 Signal

What does an MFSK16 signal consist of? Well, there are 16 tones, sent one at a time at 15.625 baud, and they are spaced only 15.625 Hz apart. To put this tone spacing in perspective, a "narrow" PSK31 signal is more than 31 Hz wide — twice as wide as the space between just two MFSK16 tones.

Each tone represents four binary bits of data. The whole transmission is 316 Hz wide, which is a bit narrower than a RTTY signal. And like a RTTY signal, an MFSK16 signal easily fits within the passband of a 500 Hz CW filter.

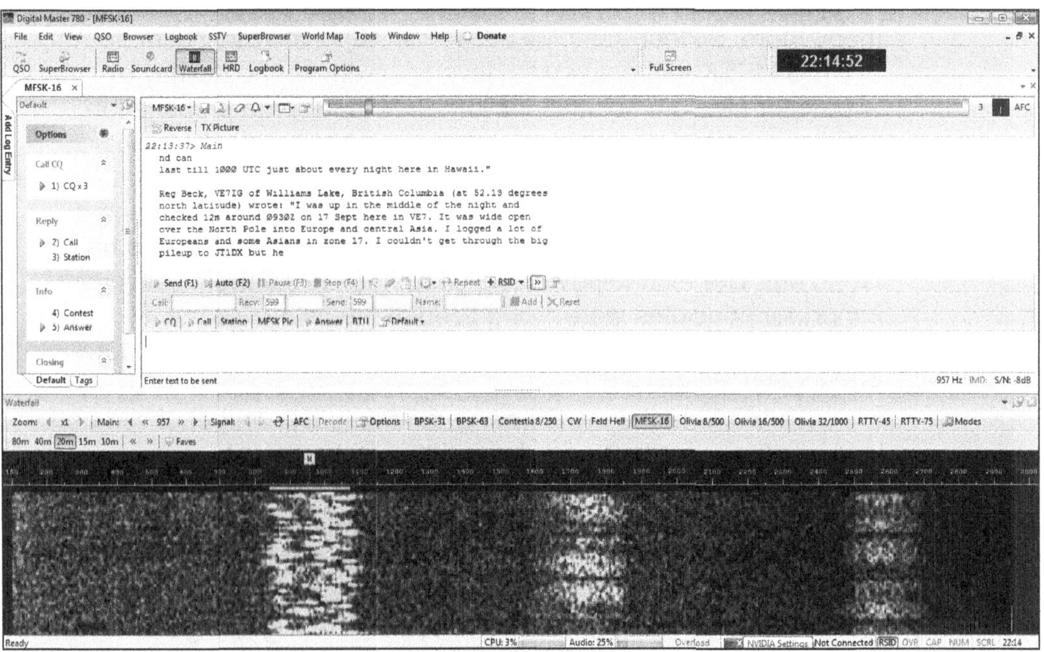

Eavesdropping on an MFSK16 chat.

When you hear an MFSK signal for the first time, you'll notice its peculiar musical quality. Some compare it to the sound of an old-fashioned carnival calliope.

Like RTTY, the signal is constant amplitude — it does not require a linear transmitter to maintain a clean signal. Driving the transmitter too hard on MFSK16 will *not* make the signal any wider. However, keep in mind that the MFSK16 duty cycle, like RTTY, is high. This can be hard on your transceiver if you are running at full RF output.

To ensure that text is received with an absolute minimum of errors, MFSK utilizes an excellent Forward Error Correction (FEC) technique, using Viterbi decoder routines by Phil Karn, KA9Q, and a clever self-synchronizing interleaver developed for MFSK by Nino Porcino, IZ8BLY. The typing rate, even with FEC, is over 40 WPM. As complicated and arcane as this may sound, the bottom line is that MFSK16 can be successfully decoded under truly dreadful conditions.

On the Air with MFSK

MFSK16 is most often heard just above the same frequencies used by PSK31. Twenty meters is the most popular band by far. If you opt to try 20 meters first, listen between about 14.072 and 14.076 MHz with your transceiver in the USB mode.

Depending on the software you are using, you may discover that tuning an MFSK16 signal takes a certain amount of patience. Because the tones are closely spaced and the software decoding filters very narrow, you must have a stable transceiver and you must use the tuning provided with the software (in the waterfall display). In the waterfall you'll notice a line that occasionally appears along the left-hand side of the column of dots that make up the MFSK16 signal. This is the *idle carrier*, the lowest of the 16 tones. This carrier is transmitted briefly at the start and end of each transmission, or whenever the operator stops typing. If your software uses a "tuning bracket" in the waterfall, you need to move the left-hand edge of this bracket until it is on top of the idle carrier line.

Each time you nudge the bracket, wait for several seconds to give the software time to synchronize and begin decoding. If the bracket isn't in the correct position, you'll see gibberish text or nothing at all.

Once you've found the right spot, almost perfect text will start to appear on the screen, although it is delayed by some 3 – 4 seconds as the data trickles through the error correction system and appears one or two words at a time.

Performance

For contacts over short paths, say distances to 8000 miles on 20 meters, MFSK16 works fine, but you may find PSK31 easier to use. If you are interested in low-power QRP operating, there is not much to choose between the two. On long path, over the poles, and in difficult conditions where interference or instability is the major problem, MFSK16 is in a class by itself. It just keeps going, giving almost perfect copy when the signal is barely audible. High power isn't necessary.

Although not noticeably better on the low bands, MFSK8 is great to have when the band starts to die. It is more sensitive than MFSK16, and although tuning is very tight, it will allow you to complete the conversation with almost perfect copy.

With MFSK16 is may also be possible to swap more than text. Some multimode programs will allow you to send tiny *images* within an MFSK16 conversation. For example, *MixW* for *Windows* software supports this image function very smoothly. If you're chatting in MFSK16 and the other station begins sending an image, *MixW* will automatically open a small auxiliary window and display the picture. If you're a *MixW* user, you can also create a macro to send an image of your own. Right click your mouse on one of the *MixW* macro buttons that you rarely, if ever, use and clear out whatever text is there. Now insert the following text:

<PIC?N%C>

Give the button a label you'll recognize such as "MFSK Pix." When you are transmitting in MFSK16, click this button and you'll be prompted to select an image file from the *MixW* directory on your hard drive (put some favorite images there for times like these). Just click the file name, then **OPEN**. *MixW* will immediately begin sending the image to the other station.

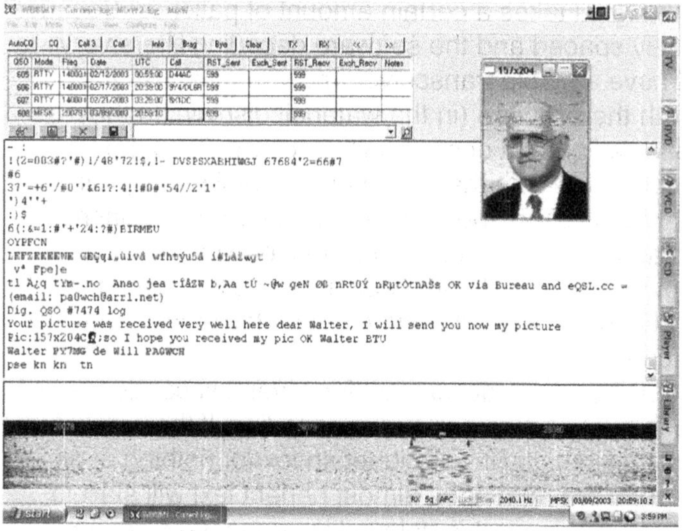

Some multimode software applications support sending small images using MFSK16.

Appendix

Steve Ford, WB8IMY, *On the Air*, July/August 2021

Winlink: The HF Email Connection

The internet has become the message information medium of choice for most hams. But there is a sizeable group of amateurs who often travel beyond the reach of the internet. This group includes hams at sea, travelers in recreational vehicles (RVs), missionaries, scientists, and explorers, not to mention amateurs active in public service.

No doubt the day will come when wireless, affordable internet access will be available from any point on the globe. Until that day arrives, however, the Winlink amateur radio HF digital network is a very capable substitute!

The Evolution of Winlink

Dozens of digital stations worldwide have formed a remarkably efficient internet information exchange network, including email and binary file transfers. Running *Winlink* software and using mostly PACTOR II, III, IV, *VARA* and other modes, these facilities transfer information between HF stations and the internet. They also share information between themselves using internet forwarding. See **Figure 1**.

An HF digital operator at sea, for example, can connect to a Winlink participating network station using *Winlink Express* software and their PACTOR controller and exchange internet email with non-ham friends and family. The sailor can also exchange messages with other amateurs by using the Winlink network stations as a traditional

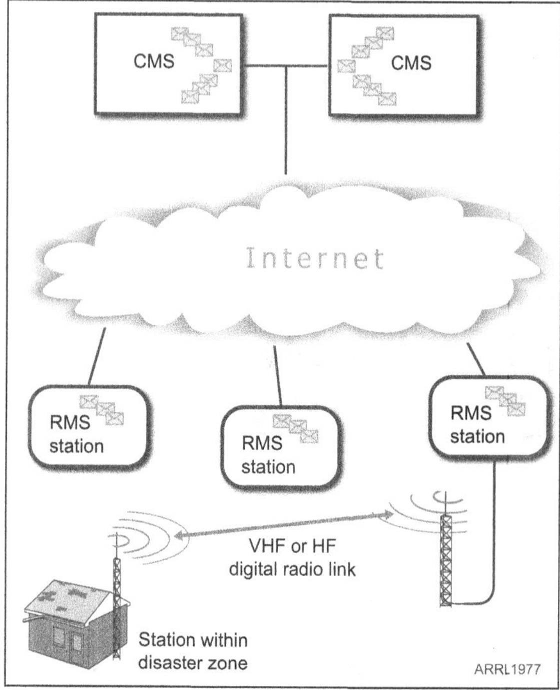

Figure 1 — A simplified diagram of the Winlink network. Common Message Servers (CMSs) store all email traffic. They share access with every RMS (Radio Mail Server gateway station) via the internet. No matter where you are located, if you can access an RMS gateway, you can pick up email addressed to you.

global "mailbox" operation.

Most Winlink participating stations scan a variety of HF digital frequencies on a regular basis, listening on each frequency for about two seconds. By scanning through frequencies on several bands, the Winlink stations can be accessed on whichever band is available to you at the time.

Using *Winlink Express*, Winlink features include:

• Text-based email with binary attachments such as DOC, RTF, XLS, JPG, TIF, GIF, BMP, etc.

• Position inquiries accessible from both the internet graphically via APRS and YotReps, email, or radio to track the mobile user.

• Graphic and text-based weather downloads from a list of over 400 weather products, covering the entire globe.

• Pickup and delivery of email regardless of the participating station accessed.

• End-user control of which services and file sizes are transmitted from the participating stations, including the ability for each user to re-direct incoming email messages to an alternate email address.

• The ability to use the internet via Telnet instead of a radio transmission.

• The ability to use any web browser to pick up or deliver mail over Winlink.

Hardware and Software Requirements

In addition to your HF SSB transceiver, you'll need a PACTOR controller, or a computer running a compatible software modem such as *VARA*.

Regardless of whether you are using an external PACTOR

controller or a software modem and a transceiver interface, you must use *Winlink Express* software to access the Winlink network. You can download *Winlink Express* free of charge at **www.winlink.org**. It contains a spell-checker and has the look and feel of an email "agent" such as *Outlook Express*.

Winlink Express will handle all the communication. If your transceiver is capable of computer (CAT) control, *Winlink Express* will even set your frequency automatically to match the frequency of the Winlink radio gateway station you are trying to reach.

Accessing a Winlink Station

Information about getting started with Winlink may be found at **www.winlink.org**. Click on the BOOK OF KNOWLEDGE menu button.

In a nutshell, once you've accessed a Winlink gateway station you use *Winlink Express* to compose your message just as you would in internet email. When you click the SEND button, *Winlink Express* will access your PACTOR controller or software modem and attempt to upload to the Winlink station. If the connection fails, it will try again later or give you the option of choosing another station.

Winlink Express once again behaves just like an internet email client. It will upload your message and then "ask" the Winlink station if you have any mail waiting for you. If so, it will download your mail automatically.

Remember that Winlink stations usually scan through several frequencies. If you can't seem to connect, the Winlink station may already be busy with another user, or propagation conditions may not be favorable on the frequency you've chosen.

Sending Email to and from the Internet

From the internet side of Winlink, friends and family can send email to you just as they would send email to anyone else on the Net. In fact, the idea of Winlink is to make HF email exchanges look essentially the same as regular internet email from the user's point of view. Internet users simply address their messages to **<your call sign>@winlink.org**.

Through the *Winlink Express* software the ham user can address messages to non-hams, or to other hams, for that matter, by using the same format used in any other email program.

Although Winlink supports file attachments, remember that the radio link is slow (especially compared to the internet). Sending an attachment of more than 40,000 bytes is usually not a good idea. Text, RTF, DOC, XLS, JPG, BMP, GIF, WMO, GREB and TIF files are permitted if they are small enough to comply with the user-set limit. Where possible, Winlink compresses files. For example, a 400,000-byte BMP file may be compressed to under 15,000 bytes. However, this is an exception rather than the rule.

Steve Ford, WB8IMY, *QST*, August 2016

The Flavors of PSK

Once upon a time, there was only Binary Phase Shift Keying at a signaling rate of about 31 baud — far better known these days as simply PSK31. Yes, Quadrature Phase Shift Keying (QPSK31) debuted at the same time, but despite the fact that it offered Forward Error Correction, it never gained traction in the HF digital community. PSK31 grabbed the crown early and hasn't relinquished it since.

Even so, the Phase Shift Keying modes have undergone an interesting evolution in the 16+ years since their debut. There are faster versions, such as PSK63, which is starting to enjoy an uptick in popularity. But as speed increases, so does bandwidth. As you'll see in **Table 1**, when you push the throttle up to 500 baud with BPSK500, the effective bandwidth increases substantially.

Another issue with PSK is that it doesn't always work well when the going gets tough. It doesn't stand up well to interference, and polar flutter can render it useless. One could switch to highly robust HF digital modes such as Olivia, but not only does the bandwidth tend to increase, the intense software processing can slow…a conversation… to…a…crawl.

Table 1
The Many Flavors of Phase Shift Keying

Mode	*Symbol Rate*	*Effective Bandwidth (at -26 dB)*
BPSK31	31.25 baud	62.5 Hz
BPSK63	62.5 baud	125 Hz
BPSK125	125 baud	250 Hz
BPSK250	250 baud	500 Hz
BPSK500	500 baud	1000 Hz
QPSK31	31.25 baud	62.5 Hz
QPSK63	62.5 baud	125 Hz
QPSK125	125 baud	250 Hz
QPSK250	250 baud	500 Hz
QPSK500	500 baud	1000 Hz
PSK63F	62.5 baud	125 Hz
PSK125R	125 baud	250 Hz
PSK250R	250 baud	500 Hz
PSK500R	500 baud	1000 Hz

Robust PSK

Developers have tried to strike a compromise by creating so-called "robust" versions of PSK modes. These are somewhat similar to MFSK in that they rely on convolutional encoders and interleavers, as well as "soft bit" decoding to increase the odds of decoding correct sequences. The result can be a fairly dependable link, despite the fact that the data rate (not the signaling rate) is cut in half compared to the standard BPSK modes.

The robust PSK modes that I've seen to date have been mostly used in data transfer applications, such as with PSK125R (the "R" stands for "robust") where it works remarkably well. The downside is that your transceiver has to operate in a very linear fashion. I've long harped on the need to avoid overdriving your radio when operating HF digital, but with robust PSK it is critical. An even slightly overdriven signal can result in poor decoding at the receiving end.

PSK63F

For keyboard-to-keyboard chats under difficult conditions, I've been taking a closer look at a variant of robust PSK known as PSK63F. The "F" in PSK63F stands for "Forward Error Correction." It is

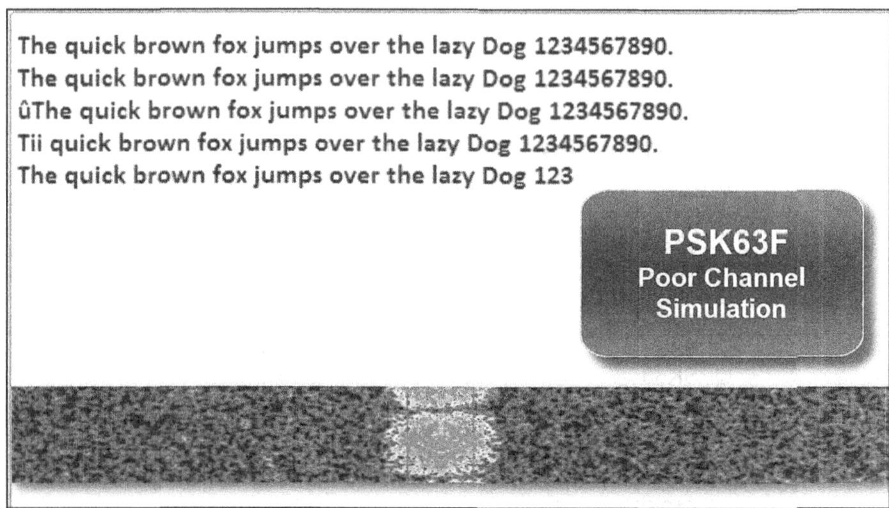

Tony Bombardieri, K2MO, posted this PSK63F test video on YouTube at www.youtube.com/watch?v=SyXQQc-4vNU. Notice the almost perfect decoding despite the simulated poor conditions.

almost identical to the other robust PSK modes except that it does not include an interleaver.

PSK63F offers the speed of "normal" PSK63, but adds a remarkable ability to be decoded under poor conditions. In my dabblings with PSK63F, I've noted that it seemed quite resistant to static crashes, which is good during the summer months. Moreover, I have also noticed that it is superior to PSK63 (and PSK31, for that matter) on polar paths.

Tony Bombardiere, K2MO, has discovered similar performance advantages to PSK63F and he has been trying it on 40 and 80 meters this summer. To demonstrate the advantages of PSK63F, he put together a short video that shows how the mode performs under simulated HF channel conditions. According to Tony, error rates have been as much as 50% lower with PSK63F compared to BPSK and QPSK. You can watch the video online at **www.youtube.com/watch?v=SyXQQc-4vNU**.

I'm not advocating a mass switchover to PSK63F, but if you're in an experimental mood, it is worth a try. At the time of this writing, I've only encountered two multimode software applications that include PSK63F: Fldigi (www.w1hkj.com) and MultiPSK (**http://f6cte.free.fr/index_anglais.htm**). Both applications are available free of charge, and both support RSID to help others identify the mode you are using. Activate RSID and try some PSK63F CQ calls. You may be surprised at who answers.

Steve Ford, WB8IMY, *QST*, September 2021

Band Hopping with WSPR

The Weak Signal Propagation Reporter, better known as WSPR, is one of the digital modes available in the *WSJT-X* software suite. Over the years, it has become the gold standard for antenna testing and propagation research. It will become an even more valuable tool as we experience the changes anticipated during Cycle 25.

On any given day, you'll find hams transmitting low-power WSPR signals from various points throughout the world. Thanks to the extraordinary ability of *WSJT-X* to decode these very weak signals, and to report the results to online data aggregation sites such as WSPRnet (**www.wsprnet.org**), it is possible to observe propagation conditions in near real-time.

Of course, propagation conditions change according to frequency and time of day, among other factors. That's why some WSPR users change bands throughout the day.

If your transceiver is connected to your station computer for CAT (computer aided transceiver) control, *WSJT-X* can automatically change bands for you in steps that coordinate with other stations. The result is more useful propagation data for everyone to enjoy. This will become critical as we approach the peak of Cycle 25.

Minute by Minute, and More

Switch *WSJT-X* to WSPR mode and you'll see a checkbox labeled BAND HOPPING. Check this box and then click on the SCHEDULE button.

You'll be presented with a matrix of checkboxes (see **Figure 1**) that

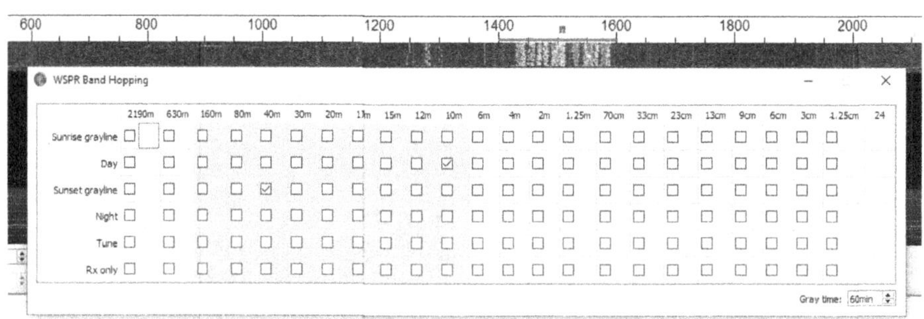

Figure 1 — The WSPR band-hopping matrix in *WSJT-X*.

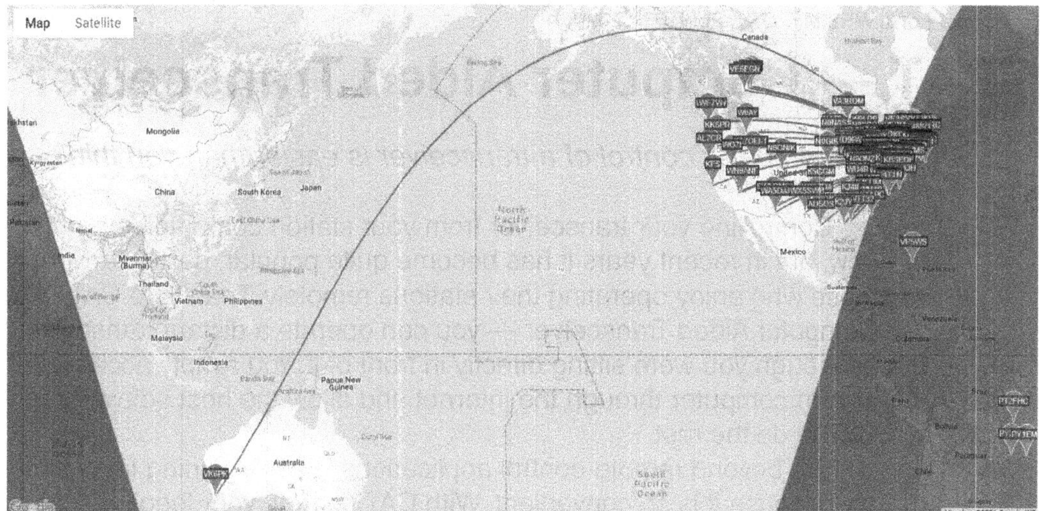

Figure 2 — VK6PK decoded WB8IMY's WSPR transmission on 20 meters while both stations were in their respective sunrise/sunset grayline zones.

allows you to select your bands of interest. You can choose to make transmissions during the day, night, at sunrise or sunset grayline periods, or you can choose to only receive during any of these periods.

WSJT-X will automatically switch your radio to each band you've selected about three times each hour. Everyone who has enabled band hopping will switch to the same band (assuming they selected it in the matrix) at the same time. If you are transmitting, the WSPR algorithm guarantees at least one transmission on each chosen band every 2 hours. The result is many stations transmitting (or listening) to specific bands at specific times, and this makes it much easier to see trends at your own station and in the online data.

One thing that's particularly clever about the band-hopping feature is its ability to calculate sunrise or sunset times for your location and change bands accordingly. In **Figure 2**, you can see a WSPRnet map indicating that Peter Hackett, VK6PK, near Perth, Australia, picked up my 5 W transmission from Connecticut on 20 meters as the sunset/ sunrise graylines swept over our respective locations. I had selected 20 meters among several bands in the band-hopping matrix and had also chosen "sunset grayline." When the program hopped my radio to 20 meters as twilight fell over New England, it sent a transmission that Peter was able to decode not long after his sunrise.

Steve Ford, WB8IMY, *QST*, February 2020

CAT — Computer Aided Transceiver

Setting up computer control of a transceiver is easier than you think.

Controlling your transceiver from your station computer is hardly new, but in recent years it has become quite popular, particularly for those who enjoy operating their stations remotely. Thanks to CAT — Computer Aided Transceiver — you can operate a distant transceiver as though you were sitting directly in front of it. You simply access the station computer through the internet and allow the host software and CAT to do the rest.

But beyond remote-control applications, CAT is gaining favor simply because it is so convenient. With CAT control, your logging software always "knows" your frequency and operating mode — all you have to do is enter the call sign of the station you contacted. If you use *WSJT-X* for FT8 and other modes, the software will use CAT to change frequencies with just a mouse click on a drop-down menu (it can even hop from band to band automatically when operating WSPR). The same is true for many other CAT-enabled applications.

Before we dive deeper into CAT, however, it is important to point out that CAT control is not necessarily the same as transmit/receive keying you might use through an interface for, say, CW or digital operating. The only "control" going on in those activities is whether the transceiver is transmitting or receiving; your computer is not communicating with the radio in the same sense that it does with CAT.

The Official CAT

The CAT acronym has become generic these days and is often used to mean any type of transceiver control. However, there is only one "true" CAT.

By definition, CAT is a control interface based on the RS-232 data standard. If you've been around personal computers for a while, the standard should be familiar. Depending upon the application, CAT is usually a point-to-point protocol: a single computer controlling a single radio. Many CAT systems rely on ASCII data and the "meaning" of a given ASCII character is particular to the brand of the radio you are attempting to control.

This means that the software must send the specific ASCII characters the radio expects to see. By the same token, the radio will respond to the software with specific ASCII characters. Think of it as your computer and transceiver being required to correctly speak the same language, although computers are not nearly as forgiving as human brains. I can fumble through Spanish and make myself understood in a simple exchange, but that's not sufficient in the computer world. Even the slightest error in "pronunciation" over a CAT line will totally disrupt the communication.

When you reach for the radio and change frequencies, for example, your radio will respond by sending data to the computer to say, in effect, "This is my new frequency." But if the data isn't what the software running in your station computer expects to see, the computer will reply with something akin to "Huh?" and display nothing or possibly even gibberish on your monitor.

Not only do your computer and transceiver need to speak the same language, they must do so at the same speed and in the same data format. The signaling rate — expressed in *baud* — must be the same. The number of *data bits* and *stop bits* must also match. This is why every piece of CAT-enabled software includes a menu in which you establish these parameters.

Matching the Data Parameters

Aside from speaking the command dialect your transceiver understands, successful CAT communication requires matching data parameters. I call these parameters the Big Four.

Baud Rate: Baud is the unit for the symbol or modulation rate in symbols or pulses per second. So, 9,600 baud means 9,600 symbols per second.

Data Bits: The number of bits used to represent one data character. Eight bits are the norm in CAT communication.

Stop Bits: In CAT communication, a *start* signal tells the receiving device (your radio or your computer) to get ready for new data. You won't be asked to specify this in your software setup. However, you *do* have to choose the number of stop bits. These bits essentially execute a reset so a new data sequence can begin. In most instances, one stop bit is all that is needed.

Parity: A simple form of error checking. When data is transferred electronically, it's not uncommon for bits to "flip" — change from a 1 to a 0, or vice versa. A parity check can detect these errors. Let's say you have a binary sequence that is using "even" parity. In that case, a parity check counts the total number of ones and if the sum is not even, that means an error is likely to have occurred. If you're asked to set parity, your choices are typically odd, even, or none at all. For CAT, the choice is often "none."

Icom CI-V

Icom's version of the CAT interface is known as CI-V. CI means "Communication Interface" and V is the Roman numeral 5 — Computer Interface Five.

CI-V is also based on RS-232, but it is a serial "one-wire" protocol. Unlike many CAT implementations, CI-V allows control of multiple radios on a single communications line. By using CSMA/CD (carrier sense, multiple access, collision detection), multiple radios can avoid using the communications interface at the same time, and they can detect when a collision occurs and will retransmit the data. If you work in Information Technology, you'll recognize this as the same technique used in 10BASE2 Ethernet.

With CI-V, each radio has an address and the computer can communicate with any of the radios on the CI-V bus by using their unique addresses. The CI-V protocol allows for tuning the radio, changing modes, selecting memory channels, and other functions. CI-V generally treats the RS-232 data as bytes and references commands using their hexadecimal values.

CAT Over TCP/IP

In recent years, CAT has also been implemented over TCP/IP (an internet standard communications protocol pair), often for use with software-defined radios. For example, in the FlexRadio Systems *SmartSDR* program, a TCP/IP port can be added that responds to CAT commands. So, the base protocol remains the same, but the communications transport mechanism is TCP/IP.

The FlexRadio *SmartSDR* approach to CAT supports integration with software that was designed for more traditional radios. This means that your logging program should communicate smoothly with a FlexRadio transceiver, but may be somewhat limited in what it can do. After all, the CAT standard didn't anticipate that we would see the day when a single transceiver would be able to receive on several frequencies simultaneously.

Connecting with CAT

Most modern transceivers support CAT control and they do it through a cable connected between the radio and your computer, or between the radio and a hardware interface that supports CAT. In years past, making the physical connections could be complicated,

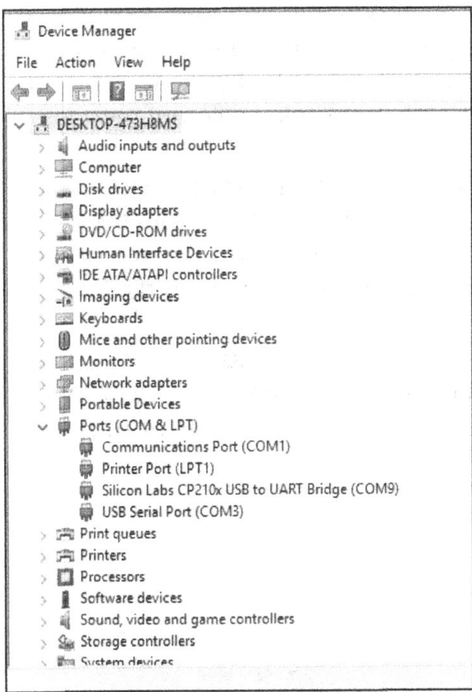

Figure 1 — Windows 10 Device Manager. This is where you can determine how Windows has assigned your virtual COM ports.

depending on the types of connectors needed. Thanks to the advent of USB, however, it is much easier today.

If you are using a USB interface that supports CAT, a single cable will connect the interface to your computer. Another cable, or set of cables, connects the interface to your transceiver. If you are blessed with owning a radio that has its interfacing circuitry built-in, you may need only one or perhaps two USB cables between your computer and radio.

In either case, when you plug in the USB cables, Windows (the operating system used in the vast majority of stations) will recognize the device — your interface or your radio — and will automatically set up one or more virtual COM ports. It will assign specific numbers to each port and you'll need to know these numbers before you attempt to configure your software.

In Windows 10, click on SETTINGS in the START menu. Now type "Device Manager" in the search window. Windows will display the search result; click on it.

Soon you will see a list of devices similar to what you see in **Figure 1.** Click the arrow beside PORTS (COM & LPT) and you will be presented with a list of all ports presently in use. The port for your radio or interface may be labeled with a name you may not recognize.
If you are unsure, unplug the USB cable and notice which port suddenly vanishes. Plug the cable back in and write down the port number when it reappears.

The image in Figure 1 was taken from my station computer. My transceiver is an Icom IC-7300. When I plugged in the USB cable between the '7300 and my computer, Windows assigned it to virtual COM port 9. You'll see it labeled in the port list as "Silicon Labs CP210x USB to UART Bridge (COM 9)."

I can use this connection to fully control the IC-7300, but I do some peculiar things with software and I needed to run CAT on a separate line. So, to meet my needs, I purchased a USB CI-V interface on eBay for about $30 and I have it connected separately. Windows assigned it to COM 3 and labeled it accordingly. You'll see it on the list as well.

Now let's turn to your software. In this example, I will use the popular *WSJT-X* application. Other programs will display different menus, but the fundamentals are the same.

Figure 2 is the *WSJT-X* radio settings menu. Look at the top of the menu and you will see that I have selected the IC-7300 from the list of radios. By doing so, I've told *WSJT-X* that it needs to use the proper language to communicate with my transceiver. In other words, I've asked *WSJT-X* to "speak" the IC-7300 CAT dialect.

On the left-hand side, you'll see an area labeled CAT Control. This is where I set the parameters necessary to communicate with my radio. From the top down, the first selection is the serial port. For me, that is COM 3.

Figure 2 — The radio setup menu in *WSJT-X*. For CAT control, we're interested in the choices on the left-hand side of the menu.

The next descending rung on the ladder is the baud rate. All transceivers that support CAT have a default baud rate — a rate that is set at the factory. Look in your manual and find the default rate. You can always change it, of course. If you are using an older radio, or a used radio, the rate may have changed. Access the data rate menu in the radio and check to be sure.

In my case, I have set the rate to 19,200 baud. That is much faster than most operators need, but once again, I have odd requirements. If you plan to do a lot of data handling over the line, such as transferring a panoramic display from your transceiver to your computer, faster is better. Your manual will usually specify an appropriate rate.

Finally, below the data rate selection, I have set 8 bits, 1 stop bit, and no handshaking (handshaking isn't needed in most CAT systems). This is the standard CAT configuration although, once again, consult your transceiver manual.

With *WSJT-X* communicating with my transceiver, I can change frequencies — and switch to the correct frequencies in each band — by just clicking my mouse on the drop-down menu (see **Figure 3**).

I've provided another example in **Figure 4**. This is the CAT menu used by N3FJP's *Amateur Contact Log* software. Notice that the COM port, the data rate, the number of data bits, and the number of stop bits are the same. *Amateur Contact Log* gives you the ability to select *parity* error checking. You'll usually set this to "none" for CAT applications.

Amateur Contact Log also allows you to set the polling rate, which is a parameter that dictates how often the software taps your transceiver on the shoulder and requests an update, so to speak.

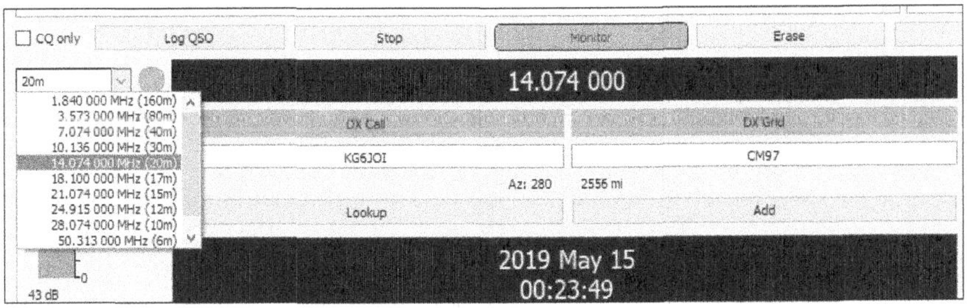

Figure 3 — With *WSJT-X* communicating with my transceiver, I can change bands and frequencies with just a click.

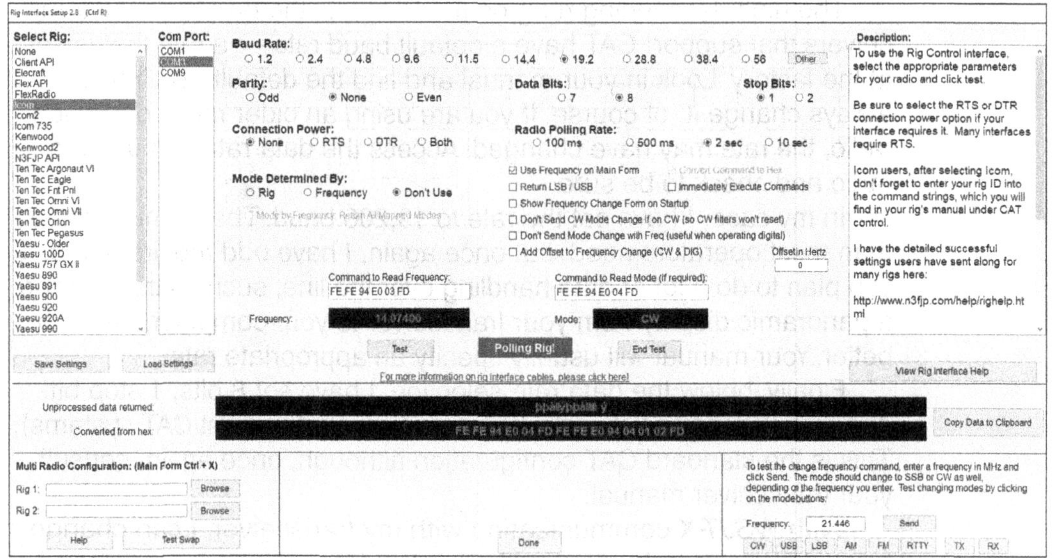

Figure 4 — The Rig Interface Setup **menu in N3FJP's** *Amateur Contact Log* **software.**

Every 2 seconds is enough for me.

The CAT is in Control

There is a certain satisfaction in seeing your operating frequency suddenly displayed in your software. And as you spin the VFO knob, the software display changes as well. Your computer and radio are carrying on a friendly conversation and all is well with the world.

CAT may seem like a luxury to some, but once you've got it up and running, I'm willing to bet you will never go back.

Howard "Skip" Teller, KH6TY, *QST*, March 2011

Digital VOX Sound Card Interface

Most sound card interfaces are powered either by a voltage from the computer serial port, by a voltage taken from the computer accessory port or microphone port, or by a "wall wart" ac adapter. If the computer has no serial port, and most computers these days have USB ports instead of serial ports, it is also necessary to use a USB serial adapter to generate a *virtual* serial port that the communications software can use for push-to-talk operation.

It would be more convenient if no dc voltage were needed to power the interface, and also if no serial port or USB serial adapter were needed, so I wanted to find a way to eliminate both the need for a serial port or USB serial adapter and a dc voltage. I then realized that computer sound cards have evolved from having both a high level speaker output jack and a line-level audio output jack, to usually having just an earphone/headphone jack and a microphone jack. Measuring the maximum audio output level of this jack on several computers, and also on an external sound card, such as "USB sound adapters" commonly sold to provide microphone and earphone jacks via a USB connector, I found that it was generally around 2.5 V peak-to-peak — not enough to power a switching transistor in an interface.

By connecting the earphone/headphone output to the center tap of a 600:600 Ω isolation transformer, however, that voltage is doubled across the full secondary winding of the isolation transformer to 5 V peak-to-peak — enough to rectify and power a transistor switch for push-to-talk switching. Since all this occurs at the secondary winding of the transformer, the transformer still isolates the computer earphone output and ground from the transceiver itself, thereby preventing any hum or ground loops from disturbing the transmit or receive audio.

By using another isolation transformer for the receive audio, the computer is totally dc-isolated from the transceiver, both on the audio input lines for transmit, the receive audio output line for receive, and the push-to-talk switching line for transmit/receive switching. The schematic diagram in Figure 1 shows how this isolation is provided by the transformers.

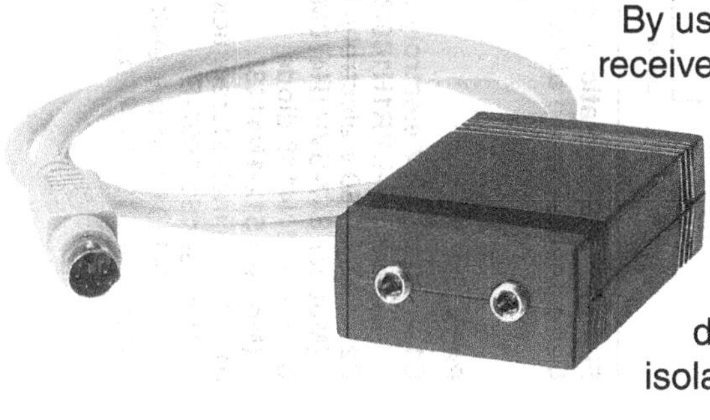

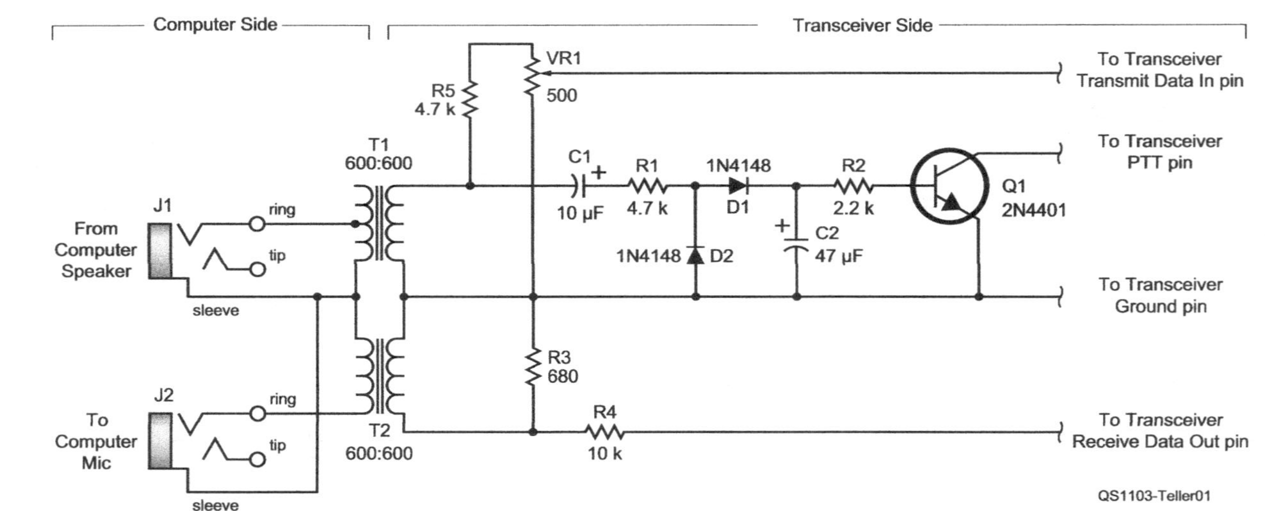

Figure 1 – Schematic diagram of the Digital VOX Sound Card Interface. Vendor part numbers are shown in parenthesis. ***Components can be obtained from Mouser Electronics, 800-346-6873; www.mouser.com. Parts list on facing page.***

Parts List

C1 – 10 µF, 50 V electrolytic capacitor (Mouser 647-UVR1H100MDD1TA).
C2 – 47 µF, 50 V electrolytic capacitor (Mouser 647-UVR1H470MED1TD).
D1, D2 – 1N4148 diode (Mouser 512-1N4148).
J1, J2 – ⅛ inch stereo jack (Mouser 161-3507-E).
Q1 – 2N4401 or other switching transistor (Mouser 512-2N4401BU).
R1, R5 – 4.7 kΩ, ¼ W resistor (Mouser 291-4.7K-RC).
R2 – 2.2 kΩ, ¼ W resistor (Mouser 291-2.2K).
R3 – 680 Ω resistor (Mouser 291-680-RC).
R4 – 10 kΩ, ¼ W resistor (Mouser 291-10K-RC).
T1, T2 – 600CT:600CT isolation transformer (Mouser 42XL016-RC).
VR1 – 500 Ω potentiometer (Mouser 652-3362-1-501LF).

How It Works

See **Figure 1**. Sound-card-based digital communications software, such as *DigiPan*, generates a WAV audio signal when placed in the transmit mode. This WAV audio contains the modulation that you use to communicate on digital modes, such as PSK31 and others, and is used to modulate the transceiver in the same way that you modulate the transceiver audio with the microphone when operating SSB. Since this WAV audio comes out of the earphone or speaker jack of the computer, that jack is connected to isolation transformer T1, but only between one side of the primary winding and the center tap of the primary winding. When this audio is coupled across the transformer to the full secondary winding, it appears at twice the value that is present between the primary center tap and either end of the primary winding, because the turns ratio of the transformer in that case is no longer 1:1, but 1:2.

The WAV audio is then used to modulate the transceiver through the data, accessory or microphone jacks for digital operating. The audio level is adjusted by means of potentiometer VR1.

In order to switch the transceiver from receive to transmit, this same double-value ac voltage is applied to capacitor C1 and resistor R1, which isolate the transformer audio from the switching action of the following voltage doubler circuit, D1 and D2. These diodes form a classic dc voltage doubler rectifier circuit, which conducts on both positive and negative cycles of the WAV audio. C2 then charges up to the peak value of the ac voltage and holds that charge long enough to drive the base of switching transistor Q1, through current limiting resistor R2, causing the collector of Q1 to saturate and pull the push-to-talk pin of the transceiver to ground. Q1 gets its operating collector voltage from the push-to-talk circuit in the transceiver, which must be designed, as almost all transceivers are these days, for an "open collector" switch for transmit/receive switching.

Transformer T2 is used to isolate the receive audio to the computer from the transceiver, and the primary is connected to the computer microphone input. The receive audio output voltage is fed from the transceiver through an L-pad consisting of R4 and R3, attenuating the high audio output of the transceiver data jack, or earphone jack, to a suitable level for the microphone input of the computer. It is this input to the microphone of the computer that creates the "waterfall" display of the typical digital communications program that also decodes the WAV audio being transmitted by the other station into characters on the screen. Resistors R4 and R3 can be exchanged in position if the computer audio input requires a higher audio level from the transceiver.

Adjusting the Transmit Level

Before connecting the audio cables, you should be able to hear the WAV audio in the computer speakers when the software is in the transmit mode. Plug in both cables between the computer and transceiver. With the software waterfall cursor around 1500 Hz and the software in transmit, raise the audio output level until the transceiver switches into transmit. Once again, if the software you are using doesn't offer the means to make this adjustment you'll need to access the audio levels within *Windows* Control Panel.

When the transceiver switches into transmit, increase the audio output level about one more notch to ensure there is enough audio for push-to-talk to consistently activate. Adjust the transceiver RF power control for maximum, or if using the microphone jack on the transceiver, to the normal position for SSB phone operation. Now adjust VR1 until the RF output of the transceiver is about 30% of rated maximum power. It should not be necessary to make this adjustment again. Just raise the audio output level control to obtain a little more output power if desired, but at 30% power, you should automatically have a clean PSK31 signal.

After setting the RF power level, place the wired circuit board into the enclosure and snap the two halves together. This completes the assembly and adjustment of the interface. If you find it necessary to adjust the interface very often, it might be convenient to drill a small hole in line with VR1 so a screwdriver can be inserted into the enclosure to adjust VR1. In most cases, however, it should be possible to just set VR1 and leave it alone, doing all the fine power setting adjustments with the software audio level controls. Near the extreme edges of the IF passband, the audio output of the transceiver may decrease because of the shape of the IF filter, so it may be necessary to increase the audio output level to maintain push-to-talk action. Remember to recheck the power output when retuning to the center of the passband.

The digital VOX interface works well with all digital modes such as PSK31 and AFSK RTTY, but does not work correctly with sound card CW or Hellschreiber. If you intend to operate those modes, the Classic Sound Card Interface described on page 37 of the July 2010 *QST* is a better choice because its serial port switching keeps the transceiver in transmit until the software returns it to receive. It is not dependent upon the audio tones being present to keep it in transmit.

Index

D

E

F

N

O

P

Q

R

S

T